Self, Senility, and Alzheimer's
Disease in Modern America

Self, Senility, and Alzheimer's Disease in Modern America

A History

JESSE F. BALLENGER

The Johns Hopkins University Press
Baltimore

© 2006 The Johns Hopkins University Press
All rights reserved. Published 2006
Printed in the United States of America on acid-free paper

2   4   6   8   9   7   5   3   1

The Johns Hopkins University Press
2715 North Charles Street
Baltimore, Maryland 21218-4363
www.press.jhu.edu

Library of Congress Cataloging-in-Publication Data
Ballenger, Jesse F.
Self, senility, and Alzheimer's disease in modern America : a history /
Jesse F. Ballenger.
p. cm.
Includes bibliographical references and index.
ISBN 0-8018-8276-1 (hardcover : alk. paper)
1. Alzheimer's disease—United States—History—20th century. I. Title.
[DNLM: 1. Alzheimer Disease—history—United States. 2. Attitude to Health—
United States. 3. Self Concept—United States. 4. History, 20th Century—United
States. 5. History, 19th Century—United States. WT 11 AAI B191s 2006]
RC523.B25 2006
362.196′831—dc22   2005012140

A catalog record for this book is available from the British Library.

For David D. Van Tassel (1928–2000)
Know the past, be the future . . .

Contents

Preface

The roots of this book stretch back twenty years, long before I had any idea that something like a cultural history of senility and Alzheimer's disease was possible, let alone that I would write one. Between the time I graduated from high school in 1979 and began work on a Ph.D. in history in 1992, I worked as a nursing assistant in a couple of medium-sized general hospitals. I was typically assigned "total care" cases on the geriatric wards: old, frail, chronically ill patients who needed to be helped with every aspect of their physical functioning—moving, bathing, dressing, eating, excreting—and who were more often than not at least mildly confused and in many cases in advanced stages of dementia.

During the years I worked in hospitals, Alzheimer's disease emerged from obscurity to become the dominant diagnostic category and way of representing dementia in old age. I recall an in-service training in which nursing assistants were indoctrinated in the new language of Alzheimer's disease. "There is no such thing as senility," we were taught. When aged people became chronically confused, it was not simply a matter of getting old; they in fact had some kind of disease, most commonly Alzheimer's, and should not be stigmatized as senile. When, as a graduate student in history, I looked back on this experience, it seemed that such a dramatic transition in how we understand and talk about dementia was worth examining as a historical phenomenon.

As I explored the ways in which this transition in the terminology used to describe age-associated dementia might be more broadly connected to American culture, I was eventually drawn back to my experience taking care of these patients. Working through the popular and professional literature on Alzheimer's, I became particularly interested in the issue of fear: Why was Alzheimer's so widely regarded as the worst and most devastating of all possible diseases—the "funeral that never ends"? The answer seemed to be that Alzheimer's somehow destroyed the selfhood of people with the disease, that the destruction of brain tissue entailed a destruction of abilities and qualities essential to people's identity, if not to their very humanity. People with Alzheimer's disease were typically described as "hollow shells," as "no longer really there," as having somehow already died despite the troubling persistence of an animate body.

But this idea, reiterated countless times in the great outpouring of material about Alzheimer's published since the 1970s, seemed puzzling in light of my experience as a nursing assistant. The nightmarish aspects of living in severe chronic confusion were, of course, plain and undeniable. Patients with severe dementia were often terribly frightened or angry, unable to escape an environment they perceived as perpetually strange and often hostile, unable to make sense of the steady flow of strange faces caring for them, sometimes even unable to recognize or understand what family members were doing in their rooms when they came to visit. Yet the nightmare of dementia itself seems to speak strongly for the persistence of a self with which to experience it. As a nursing assistant with the task of feeding, bathing, and moving sometimes as many as eight such patients in an eight-hour shift, I certainly could see that even people with severe dementia had an identity and an independent will that had to be reckoned with.

I recall one such patient vividly because his name was also Jesse, and I used that to try to get him to remember my name. I took care of him often over the course of two or three months in the hospital's extended care facility, until he was transferred to a nursing home. He was lean and well muscled for a man who was probably in his seventies, still able to walk fairly well and feed himself—though so confused that you either had to restrain him or had to watch him constantly to keep him from wandering away or getting into trouble. His wavy hair was silver with dark streaks, a little too long and hopelessly unmanageable. He had glasses with thick plastic frames and large, strong lenses that made his brown eyes more prominent as he stared at you from under the thick, tangled wire of his eyebrows. If you could take his craggy face out of context, you might think him a scientist or engineer—someone who was forceful and clear, perhaps brilliant, but who just couldn't be bothered with appearances.

When I would go into his room in the morning carrying his breakfast tray, he'd usually be straining to get out of the vest they tied him to his bed with at night. "Buddy," he would call, "can you give me a hand? I'm stuck here and I gotta get to work. I'm late. Can you just get me a knife?"

"Hang on, you need to have some breakfast first."

"I don't have time for breakfast. Just get me loose. Just cut this damn thing off of me, I don't care," he'd say, tugging at the vest as I knelt down to undo the straps where they were fastened under the bed.

"Oh come on, Jesse. You've got to have some breakfast before you go to work." As I helped him to his feet and guided him toward the bathroom, careful to keep the straps of the vest out of sight behind him, I'd say, "Hey, do you remember who I am? Here, let's get you into the bathroom. Do you remember who I am, what my name is? I have the same name as you."

He'd pause, stare at me, puzzled, seeming to sense from my question and tone that he *should* remember me. Clearly he had no idea who I was. But he would never admit that. "Sure I remember you," he'd say. "Sure."

In all the time I took care of him, he never could make the connection and would just repeat his own name and stare blankly at me. We'd continue on that way through the day, him trying to get free of the inexplicable and maddening barriers that kept him from getting where he needed to go, and me trying to guide him where I needed him to go and divert his attention from the fact that the vest was going to have to stay on. If I was lucky, he would resign himself to eating his breakfast and maybe fall asleep in the geri chair that I strapped him into while I took care of some of the other patients assigned to me. Sometimes, though, he'd stay angrily focused on the restraint, on his need to get somewhere, and he'd be yelling out for help, struggling to slip out of the vest and under the tray that locked into position across the chair, sometimes even throwing things from the tray.

Often, as I helped him wash, shaved him, rinsed his dentures, and got him dressed, I would play along with whatever he was telling me, no matter how far-fetched or incoherent, to try to distract him from the situation in which he actually had to live. He was usually obsessed about work. I'd ask him where he worked, whether he liked it, and so on. If he was worried about being late, I'd assure him that I could get him there on time once he got ready. Sometimes he was worried about finding a job, and I'd assure him that I could get him one. He would thank me profusely, and I'd set him up in his geri chair in the hall folding towels or linens, which he would usually find absorbing for at least long enough to let me attend to another of my patients.

It may be that his confabulations were a mosaic of his past experience, but they were so vague and jumbled that in all the time I spent talking with him I was never able to determine what sort of work he had done in his life. At odd moments, though, he would seem to have flashes of lucidity about his situation. Sometimes he seemed to be able to see through me and understand that my questions were aimed at distracting him from the fact that I was retying his vest. He'd stop talking to me and shake his head wearily, as though amazed at my pathetic subterfuge. At other times, when there was a lag in our conversation, he would stare out at me with eyes that seemed as lucid as any I have seen: "How did I get into this mess?" he'd ask. "I just don't understand how I got into this mess."

For myself, I can say that, for all the difficulty Jesse's confusion created, Jesse was genuinely interesting and fun to take care of. As with many other patients far down the road of dementia, his delusions made a kind of pleasantly absurd counterpoint to the physically demanding, tedious, and often unpleasant work, opening up a playful space in which the difficulty and boredom of the day could pass more lightly.

Moreover, playing along with Jesse M——'s delusions seemed to make him less anxious and angry. Certainly it contrasted with what often happened when his wife visited him in the afternoons, a small, very frail looking woman with wire-rimmed glasses and thin white hair pinned back neatly behind her head. Jesse recognized her only marginally better than he did me, and she seemed to find it hard not to be angry with him about this. "Don't you know me?" she'd ask in frustration and disbelief. She constantly struggled to make him understand the situation. She would hover about him, trying to get him to stop fussing with the towels, stop worrying about places he thought he needed to go. "You don't have to go anywhere," she would tell him, her voice cracking. "You're in the hospital. You're going to have to stay in the hospital. I just can't take you home like this." All of this only increased his agitation. Sometimes he would start yelling for help even as she was sitting there with him, and she could not calm him down. Other times he would tire himself out and fall asleep. She would sit next to him quietly, a crumpled tissue perpetually in her hand. His dementia was clearly such a profound emotional blow to her that trying to get him to recognize her could be no game; the practical problems and decisions she had to make were so grave and heartbreaking that there was no question of playing along with his confusion.

As I look back at such experiences now, it seems to me that the contrast between the relative ease and pleasure that I, a detached caregiver, could take in caring for people with dementia and the anguish that families often seemed to experience can explain much about why dementia generates so much fear. I found caring for patients with dementia a relatively light emotional burden because I was able to accept whatever they told me about themselves, no matter how incoherent or patently absurd, no matter how in conflict it might be with what they told me about themselves a day, an hour, or even a minute before. Put another way, my emotional distance as a paid caregiver made me willing and able to let the selfhood of these patients pass as legitimate, regardless of their inability to present a consistent and coherent account of themselves.

For family members, such acceptance was understandably much more difficult. Because their own life story and sense of identity are so deeply bound together with those of the person with dementia, the memory failures and confabulations of the latter must seem deeply threatening. I got a sense of this when my own father was dying of lung cancer during the time I was formulating these ideas. Although he remained lucid until he died, as the cancer metastasized, he began to show some mild aphasia. I was surprised how unsettling I found his occasional inability to find the right word, and I got at least some sense of the dread family members must feel as they watch someone to whom they are inextricably connected slowly lose his or her ability to recall and retell the shared story of their lives.

But the anxiety generated by dementia is not limited to the friends and family members of people with dementia. My ability to take dementia rather lightly as a nursing assistant may have been indicative of my own peculiar intellectual commitments, or simply of the distance of youth from the prospects of old age. In any case, as I have studied the public culture around Alzheimer's disease, it seems clear that dementia is one of our greatest fears. This book is an exploration of the historical roots of this fear in order to better understand the complexities of the problems Alzheimer's disease poses to our society.

A NOTE ON TERMINOLOGY

When describing popular and professional discourse from earlier historical periods, I have found it sometimes necessary to use terms such as *aged* or *senile*, which today are widely and correctly considered to be offensive and stigmatizing. My use of archaic and offensive terms at some points in this book is in no way an endorsement of such language. Contemporary usage is more accurate and appropriate. For example, the term *people with dementia* correctly recognizes that the person with dementia is not wholly defined by it (i.e., he or she is a person with an impairment rather than an impaired person) and does not imply that most older people are or will become cognitively impaired.

But in describing ideas and beliefs about aging and dementia common among physicians and the general public in earlier historical periods, I have found it necessary to repeat the terms they use in order to capture the meaning and flavor of their discourse. Quite simply, the distinctions suggested by the term *people with dementia* were not recognized in earlier historical periods, and it is confusing and inaccurate to use them when describing the beliefs of people in these periods. Moreover, we have not left this legacy completely behind, and it is important not to use delicate terminology to gloss over unpleasant realities. Doing so may

Acknowledgments

Many people in different ways have helped to make it possible for me to write this book. The Department of History at Case Western Reserve University, where this project began as a doctoral dissertation, was a particularly rich place to engage in this sort of inquiry, and in particular the mentorship of a number of faculty members at Case has been crucial. When I entered the program, David Hammack urged me to pursue research related to my experience working as a nursing assistant in the geriatric wards of hospitals, and that good advice set me on the course that ultimately led to this book. My dissertation adviser, David Van Tassel, who died shortly after I graduated from Case, was a supportive mentor, patiently guiding me through the early stages when it seemed this project would never be started, let alone completed. More than this, he was an inspiring model of professional integrity and graceful collegiality for which I am grateful and to which I shall always aspire. Jonathan Sadowsky has been an important intellectual influence throughout this long process, always willing to hear my ideas out, urging me to sharpen and push them further, and pointing to fruitful connections with broader scholarly discourses. Alan Rocke's careful reading and critique helped sharpen my thinking on many points. Beyond the Department of History, indeed beyond the disciplinary boundaries that constrain most scholars, Peter Whitehouse's mentorship has been integral to bringing this project to fruition. Through our collaborative work, he provided me with tremendously valuable opportunities to creatively explore this topic in interaction with some of the leading figures in various aspects of Alzheimer's disease research. Beyond this, his spirit of critical inquiry into the past and possible future of this disease and deep understanding of the significance of dementia as a human problem have been a touchstone at every step of the way.

An Andrew W. Mellon dissertation fellowship provided some financial support at a crucial point. Much more important, it gave me the opportunity to interact with a group of dedicated young scholars from around the university through the Mellon fellow seminar, organized by Martha Woodmansee and Jonathan Sadowsky. The supportive community of many of my fellow graduate students at Case was also instrumental in the formative stages of this project. The members of the Department of History's Works-In-Progress group read early drafts of several chapters and provided many helpful suggestions and much encouragement. When I was ready to quit, Tasslyn Frame, Dan Kerr, Bernie Jim, and everyone in the "Graduate Burial Society" reawakened my faith in the larger purpose of this endeavor and the ideals of the academy that we must struggle to keep alive. Marty Gibbons and Sue Horning provided encouragement late in the game. Most of all, I would like to thank Pat Ryan for helping me think through this project at every step of the way and for spurring me on when the going got toughest.

I am very grateful to Randall Packard and the Institute of the History of Medicine at the Johns Hopkins University for the support of a generous post-doctoral fellowship. Faculty and graduate students at the institute provided an inspiring community of dedicated scholars and offered many helpful ideas and suggestions on portions of the book. In particular, I would like to thank post-doctoral fellows Laura McGough and Gretchen Krueger for their support, and Gretchen in particular for reading late drafts of several chapters. I am also thankful to Constantine Lyketsos and Peter Rabins of the division of geriatric psychiatry and neuropsychiatry at the Johns Hopkins School of Medicine for their interest in the work of a historian and their willingness to share their time and experience.

Jackie Wehmueller of the Johns Hopkins University Press has been a wonderful editor to work with and has provided helpful advice throughout this project. Anonymous reviewers for the press provided comments on both the proposal and final manuscript which helped me improve the book in many ways.

I have not understood until now why this disclaimer is so often made, but when I look back at the collective wisdom of all those who have read parts of this book in light of my own inability to integrate fully their good advice, I feel compelled to state the obvious: the shortcomings that remain are wholly my own.

I thank my new colleagues at Pennsylvania State University's Science, Technology and Society Program, and in particular the program's director, Martin Pietrucha, for their understanding and indulgence as I completed this work.

Finally, I am grateful for the support of family and friends. My parents died before I could finish this project, but their support, encouragement, and example remain an important resource. My sister Mary stepped in to handle family matters that could not be set aside and so allowed me to stay on track. My children

Jesse and Hannah's knack for blessedly timed distraction has helped me to keep things in perspective. Thanks also to Pat and Julie Ryan and their beautiful children for so much love and friendship during this journey. Most of all, I am grateful for the love and support of my wife, Barbara. Far more than anyone, she shared with me the burdens and joys of this work and helped me fit it into the larger weave of life.

Self, Senility, and Alzheimer's Disease in Modern America

Introduction

This is not another book about scientific progress in unraveling the mystery of Alzheimer's disease; nor is it an analysis of particular ethical or policy issues, and still less a compendium of advice and practical tips for the caregiver or professional on managing dementia. There are many, many of these already. Rather, it is a book about the peculiar dread that dementia generates in American society. It examines the historical origins of this dread, its connection to broader social and cultural developments in American history, and the way in which this dread has helped to shape knowledge about dementia, health policy, and the experience of caregivers and people with or at risk for dementia. My hope is that, in addition to providing a historical account of Alzheimer's that will interest historians and other scholars, it will help people who are concerned with the disease to ask new questions as they approach the many books about the science of Alzheimer's, ethics and policy, or clinical management and caregiving.

Perhaps the most eloquent evocation of the dread that surrounds dementia appeared in a little-known novel by Wisconsin author Robert Gard called *Beyond the Thin Line*, published in 1992. Gard characterized the book as "a fictional synthesis of one of the great human dramas and tragedies of our time—Alzheimer's disease," which he based on his experience of watching a longtime friend develop dementia.[1]

Gard had written some forty books before this novel, most of them dealing with Wisconsin life, history, and culture. As described on the dust jacket of the book, Gard at age 82 seemed an excellent example of what gerontologists describe as successful aging. "Never even thinking of retiring," he maintained "a heavy schedule of writing, teaching and lecturing." Yet Gard was clearly deeply frightened of the prospect of dementia. In a passage describing the subtle early indications of dementia in his friend, Gard vividly describes his own fears of dissolution. Observing the residents of a nursing home to which his friend has been confined, Gard senses that they have all "passed some invisible line. They can never come back and their families know this and perhaps have witnessed it happening over years, but they could not prevent, nor could they identify exactly when their loved one crossed the line of no return." But where was this line of no return? How and when did one cross over it? "The line is a mystery. Little by little it becomes knowable to friends and family, but hardly ever to the stricken persons. Often they are unaware of the presence of such a line. My friend Harry didn't know, but his friends became slowly aware, very slowly, and in some disbelief. They didn't understand." In recounting various incidents in which he becomes fully aware of his friend's cognitive deterioration, Gard writes, "I could only feel shocked, and conferring with other close friends of Harry I learned that it was all true. Little by little Harry had been slipping toward the line."

For Gard, finding this line was crucial, for the line defined the boundary between the normal and the pathological, between a coherent, stable self and the incoherent, chaotic dependency of dementia. Yet the line was almost impossible to detect, except when it had been irrevocably crossed. It could be discerned only through the closest scrutiny of the behavior not only of others but of oneself as well:

> I found myself beginning to watch for the line in several older friends and even in myself. I began taking careful heed of where I put things, of noting what my daily habits were and if I ever varied from them unknowingly. Sometimes I thought I could discern small lapses—at home I would head for another room to get a certain object and entirely forget what it was I went to fetch. I learned that these small signs were almost universal in older people, however, and that the larger lapses, the confusion, the wandering away were more serious. Of these I was not guilty, and I noted that in my public addresses, of which I gave a fair number, my train of thought was never broken or inextricably lost. I could talk for two hours without notes and never lose track of where I was. I was told that this was encouraging, and that I was certainly not yet approaching the line. But the line became an antagonist. In my imagination it grew almost into a living thing, a reality, and I fancied I saw it often drawn for this person or that. When one considers the line it becomes easy to think in negative terms, and to become fanciful about many aspects of life. Often I heard friends say, "The thing I dread most is becoming senile," and I wondered whether they too were aware of the line.[2]

Although Gard felt reassured that he was not "guilty," that he was not approaching the line beyond which his status as a responsible, respectable person would be open to serious question, that reassurance could never be more than provisional and temporary. Despite being an author of more than three dozen books, despite maintaining a fully active and productive life into his eighties, Gard continued to scrutinize his behavior and that of people around him, trying desperately to detect the line of no return. But why was this so important to him? Even if its signs were detected early, nothing could be done to prevent Alzheimer's disease. In paying such careful attention to the signs of approaching dementia in others and himself, Gard was trying to reassure himself that his own selfhood was secure and stable.

In this book, I treat the dread surrounding dementia as a historical problem. When did this fearful line Gard describes dividing people with dementia from the rest of us appear in American culture? Was it simply the product of the attention given to Alzheimer's disease, or does it have deeper historical roots? Why did it appear, and with what consequences?

History adds an important perspective to our understanding of Alzheimer's disease. As I have pointed out elsewhere, it is ironic that our public discussion of a disease that robs individuals of their memory proceeds with so little appreciation of its past.[3] Without a sense of history, without the ability to construct a coherent narrative linking the present to the past as well as the future, public discourse on Alzheimer's will itself be confused, disoriented. I believe that an understanding of the history of this condition and our attitudes toward it can help us respond with wisdom and compassion.

This book concerns the origins and development of the dread that Americans express about aging and dementia; it concentrates on developments from the late nineteenth century to the present, for I argue that it was only in the late nineteenth century that the fear and anxiety we feel about dementia today clearly emerged in American culture. But before beginning to tell that story, I want to briefly address an objection that I think many readers, unfamiliar with approaching dementia as a historical problem, will have: that a decrepit old age in general and dementia in particular have always and universally generated fear, anxiety, and hostility toward the aged. In the remainder of this introduction I want to consider at some length an example from the late eighteenth century to demonstrate that, in the American context at least, the prospect of aging in general was not viewed in this earlier period with the same sort of dread that we have come to think of as natural and that dementia in particular was not regarded as the worst of all possible fates to befall a person entering old age. One might suspect that this simply reflected a lack of experience with the disease, since relatively few people would have had direct experience with dementia in a society in which an old age remained a relatively rare attainment. But even in a medical text based on direct observation of older people, some of them evidently in various stages of dementia, the different attitude toward this phenomenon is striking.

A pamphlet on old age and its infirmities published in 1793 by Benjamin Rush, the preeminent physician and a leading citizen of the young United States, was perhaps the first American medical text to include a description of what we today would be inclined to label senile dementia or Alzheimer's disease. In it, Rush noted that "the *memory* is the first faculty of the mind which fails in the decline of life. While recent events pass through the mind without leaving an impression upon it, it is remarkable that the long forgotten events of childhood and youth are recalled and distinctly remembered." In this passage, Rush appears to have been describing what twenty-first-century neuroscience would call anterograde amnesia, one of the most prominent and frustrating symptoms of early-stage Alzheimer's, caused by the destruction of neurons in the hippocampus—the structure of the brain in which the disease's pathology first appears. Elsewhere in the essay, Rush seems to be describing other aspects of what we today recognize as Alzheimer's; for example, his description of a woman between 80 and 90 who suffered such a "total decay of her mental faculties as to lose all consciousness in discharging her alvine [bowel] and urinary excretions" seems to be a description of the end stage of the disease.[4]

As these quotations suggest, elements of Rush's description resonate with contemporary ideas and concerns about Alzheimer's. But read in historical context, Rush's essay suggests that old age and dementia had very different meanings for Americans at the end of the eighteenth century. Prior to the mid-nineteenth century, discourse on old age, including medical discourse, was primarily concerned with the relationships that the aged individual maintained with God and the community of faithful. In this discourse, the physical and mental deterioration of old age appears to have generated less fear and anxiety than it would have in late-nineteenth-century America. In a society ordered by principles of hierarchy and reciprocal obligation, disease and dependency, however painful and burdensome, could also be seen as salutary illustrations of humanity's ultimate dependence on God. In the colonial period, representations of old age were rooted in religion, and the clergy remained the authoritative voice on old age—at least in New England.[5] In the view of Puritan divines, the losses of old age served as a new invitation for Christians to witness and participate in the balance and harmony of God's creation. While the physical and mental deterioration of old age entailed a painful and often humiliating loss of physical and social power for the elderly, it did not vitiate, even when the aged developed dementia, their fundamental role in the community—to acknowledge dependency and longing for the salvific grace of God.[6] As one minister put it, piety in the old hid the "imbecility of body and mind" that so often afflicted them; it gave meaning to their lives "when they would otherwise be useless and burdensome." Long after the days of productive labor were over, the elderly could still serve God, their families, and their communities through their piety, acting as "visible monuments of sovereign grace."[7]

This view of aging did not substantially change as Enlightenment ideals of rationality and human agency challenged Calvinist theology in the eighteenth century. An appreciation of natural law and human agency led rational humanists of this period to view the loss and suffering in old age as inevitable and natural evils of existence that could be assuaged and ameliorated but not eliminated. Ultimately, one simply had to accept deterioration and even death as part of the benevolent design of nature.[8]

While this discourse on aging might be seen as "positive," it should not be taken as evidence for a "golden era" in which the aged were unproblematically venerated in early America. There is ample evidence of tense generational relations during this period, and the ideal of respect for the aged did not apply equally across lines of class, gender, and race.[9] Moreover, this ideal of aging carried with it a strong obligation for the elderly to move aside. For example, in 1726 Cotton Mather instructed older folks to "be so wise as to disappear of your own accord, as soon and as far as you lawfully may. Be glad of a dismissal from any post, that would have called for your activities."[10] In 1792, Benjamin Rush expressed similar sentiments in the language of Enlightened republicanism. In an entry in his commonplace book that was evidently the basis for an essay or talk he was preparing on the virtues of death, he noted that death "relieved children from parents who kept them too long out of their estates" and, more broadly, "the world of old men who keep the minds of men in chains to old prejudices. These men do not die half fast enough. Few Clergymen, Physicians, or Lawyers beyond 60 do any good to the world. On the contrary, they check innovation and improvement."[11]

Nonetheless, it is clear that Americans in the eighteenth century had a language that could make sense of and was relatively tolerant of the difficulties of aging. Neither the piety of the Calvinist clergyman nor the rational acceptance of aging as a "natural evil" of the republican deist could avert the physical and mental deterioration of old age or guarantee that wealth and power could be maintained. But these ideals of aging did give meaning to suffering. In this discourse the senile aged, even those with deep dementia, remained fully human and connected to the community.

All of this can be seen in Rush's essay on old age, which aimed to describe the factors favoring longevity and the phenomena of body and mind that attend to old age and to enumerate the particular diseases that afflict old age and appropriate remedies for them.[12] Rush's essay was based on observations he made over a five-year period on an unspecified number of men and women over the age of 80 and also drew from well-known episodes in the lives of eminent men, such as Jonathan Swift.

For today's reader, perhaps the most striking feature of Rush's description of old age is its lack of emphasis on what is to many of us the most distressing age-related deficits—loss of memory and intellect. Rush evidently regarded the deterioration of mental faculties as a natural, if not inevitable, aspect of the aging process, for he described it not in the section on diseases that afflict old age but as one of the phenomena of body that accompany it. However, in Rush's view, the normal deterioration of old age, in contrast to deterioration produced by the effect of disease or accident, left the mind intact until the end. "Death from old age is the effect of a gradual palsy. It shews itself first in the eyes and ears in the decay of sight and hearing—it appears next in the urinary bladder, in the limbs and trunk of the body, then in the sphincters of the bladder, and rectum, and finally in the nerves and brain, destroying in the last, the exercise of all the faculties of the mind." Rush thought that few people actually lived long enough to die this way. Rather, it was disease that "generally cuts the last threads of life."[13]

Rush roughly followed this basic course of deterioration in his enumeration and description of the various phenomena he had observed in old age. He began with a number of the physical and sensory peculiarities that accompany old age—heightened sensitivity to cold, heightened receptivity to aural stimulation,[14] increased appetite, abnormal pulse, and so forth. Loss of short-term memory, "the first faculty of the mind which fails in the decline of life," was listed as the sixth of twelve such observations. Significantly, Rush did not regard memory loss, which he thought inevitable and universal in old age, as particularly alarming, for it was not necessarily an indicator of failing "understanding" or intellect, and still less an indicator of failing moral and religious faculties.[15]

Rush thought the preservation of intellect, despite even severe loss of short-term memory, was directly related to the extent to which the mind was exercised. Here Rush revealed his commitment to the Enlightenment ideals of rationality and human agency, for he contrasted his belief in the ability of old people to preserve their mind through exercise not with a belief that the elderly inevitably developed dementia but with the idea that they preserved their intellectual and moral faculties solely through the grace of God. "I have observed some studious men to suffer a decay of the memories, but never of their understandings," Rush asserted. Literary men, thanks to their "constant exercise of the understanding," often preserved it into extreme old age whatever the state of their memory. Likewise, Rush thought that older people who lived in cities preserved their intellect longer than those in rural settings because of their immersion in a more intellectually stimulating environment. Citing a well-known example that seemed to run against this explanation, that of Jonathan Swift, "one of the few studious men, who have exhibited marks of a decay of understanding in old age," Rush endorsed Samuel Johnson's view that Swift ruined his mind by "a rash vow which he made when a young man never to wear spectacles" and "a sordid seclusion of himself from company, by which means he was cut off from the use of books, and the benefit of conversation, the absence of which left his mind without its usual stimulus—hence it collapsed into a state of fatuity."[16] In contrast to Swift, Rush described the exemplary attitude displayed by author, educator, and philanthropist Anthony Benezet, who in his old age showed that failure of memory ought to excite no anxiety. Without irony, Rush reported that "even this infirmity did not abate the cheerfulness, or lessen the happiness of this pious philosopher, for he once told me, when I was a young man, that he had consolation in the decay of his memory, which gave him a great advantage over me. 'You can read a good book (said he) with pleasure but once, but when I read a good book, I so soon forget the contents of it, that I have the pleasure of reading it over and over; and every time I read it is alike new and delightful for me.' "[17]

Rush also thought that the moral and religious faculties would remain intact as memory and even understanding deteriorated. "I did not meet with a single instance in which the moral or religious faculties were impaired," he asserted simply. "I do not believe, that these faculties of the mind are preserved by any supernatural power, but wholly by the constant and increasing exercise of them in the evening of life. In the course of my inquiries, I heard of a man of 101 years of age, who declared that he had forgotten every thing he had ever known, except his God. I found the moral faculty, or a disposition to do kind offices, to be exquisitely sensible in several old people, in whom there was scarcely a trace left of memory or understanding."[18] Whatever suffering the loss of intellect involved, it did not entail a forfeiture of the most basic human qualities. Rush drew no sharp line to divide people with dementia from the rest of humanity even when describing a woman who suffered such complete destruction of her intellect as to become incontinent of urine and stool. He characterized this condition as a "second infancy," a deepening of the more common "second childhood." Although indicating marginalization, the term *infancy* clearly suggested a level of connectedness to the rest of humanity, even a receptivity to human kindness and ministration missing from the term *vegetable,* which would become more typical in representations of such cases from the late nineteenth century on. However grave their losses, people with dementia retained their essential humanity in Rush's representation, remaining connected to what were then the wellsprings of human meaning—community and God.

In the following chapters, I examine medical, policy, and popular discourses on aging and dementia to trace the rising prominence of senility as a cultural category from the late nineteenth century, culminating with the emergence of Alzheimer's as a major disease entity and public issue in the 1980s.

Of course, a significant factor in the rising prominence of dementia in the United States has been the aging of the population. In 1900, a little more than 4 percent of the total population was 65 or older; in 2000, more than 12 percent of the population was 65 or older. Although there is considerable controversy about the prevalence of Alzheimer's and related disorders,[19] widely accepted estimates for the prevalence of dementia in the United States are that about 10 percent of the population 65 and older and nearly 50 percent of the population 85 and older have dementia[20]—suggesting that the number of Alzheimer's cases has risen tremendously since the 1970s. At the same time, it should be pointed out that dementia was probably not rare in previous decades. It is highly problematic to estimate the prevalence of dementia in the past by using current percentages because the proportion of people 85 and older—who are most at risk for dementia—did not constitute as large a proportion of the elderly population as they do now and because there is evidence that people with dementia die at a higher rate than those without it, an effect that may have been more pronounced in the past. Nonetheless, since the 65-and-over population had reached more than 9 million in 1940, with 365,000 people 85 or older, and 12 million in 1960, with more than half a million of them 85 and older, it seems reasonable to suppose that the number of people with dementia in the United States may have been as high as 1 million by sometime in the mid-1940s.

Whatever the actual historical prevalence of dementia, its rising prevalence especially after 1970 tells only a small part of the story. Increasing prevalence tells us that senile dementia has attracted more attention, but it does not explain the particular forms that attention took. Why, for example, should dementia be regarded by medicine in some periods as a part of aging and in other periods as a disease? What explains the intensity of the fear and anxiety that now surround it? I argue that broader changes in American culture associated with modernity have in fact made the symptoms of senile dementia more frightening. In the ante- bellum period, hierarchical social relations eroded with the market revolution and the triumph of a liberal social order. Since then, the notion of selfhood has grown increasingly problematic for all Americans. Selfhood was no longer an ascribed status but had to be carefully and willfully constructed by every individual.[21] Constructing a self was further complicated in the late nineteenth century by the increasing complexity and power that external social forces exerted on the individual—industrial production, bureaucratic governance, and mass consump- tion.[22] In light of these changes, the loss of the ability to independently sustain a coherent self-narrative—a loss that dementia entails—has come to be considered the most dreadful of all losses.

These changes undermined not only notions of a stable self but concepts of a meaningful old age as well. Representations of senility have been part of a broader discourse on the fate of old age in modern society—about whether the aging body and mind could possibly keep up with the frenetic pace of change in an industrial age. This image of the senile person, and especially of the senile man, has been one the most prevalent stereotypes for managing anxiety about the coherence, stability, and moral agency of the self under the conflicting demands of liberal capitalism. Thus, these images of senility have played a major role in structuring the meaning of both old age and the self in modern America. Senility haunts the landscape of the self-made man.

In post–World War II America, social scientists in the blossoming field of gerontology attacked the stereotype of the senile old man no longer able to keep up with the pace of the modern world. So-called senility, they argued, was not a problem within the aging man's brain but the product of the deprivation of meaningful roles in old age—a problem that they supposed men faced more than women because aging men faced the loss of status and income with retirement. At the same time, the pressure of an aging population on postwar labor markets made retirement at age 65 imperative. Although gerontologists opposed age dis- crimination, they also tried to make retirement palatable and less psychically damaging by grounding successful old age in images of active leisure and health. Gerontologists and their allies in government, the mass media, and advertising largely succeeded in transforming retirement from a dreaded imposition to a desirable status. But, paradoxically, doing so intensified the stigma of senility. As the meaning and purpose of old age was reduced to maintaining one's health and activity levels, disease and dependency grew even less tolerable, and the prospect of losing one's mind in old age became more frightening than ever. In the 1970s, biomedical science broke down the broad notion of senility into constituent diseases such as Alzheimer's. Since then biomedicine has made impressive strides in understanding the biological processes that produce dementia, but in so doing it has only intensified the fear of and hostility toward old age by making dementia more "real."

Thus, the line between the demented and the normal which 82-year-old Robert Gard tried so intently to detect in others and in himself emerged over the course of the twentieth century in medical and popular discourse about senility and old age. It was made salient by the increasing anxiety in American culture about the stability and coherence of selfhood in general and by changes in the meaning of old age in twentieth-century America.

By the 1980s, people with dementia were at the outer limits of stigma—characteristically represented as no longer "really there," as having somehow already died despite the troubling persistence of an animate body. Thus, in her painful memoir of a troubled mother-daughter relationship, Julie Hilden wrote that she could not remember on what day exactly her mother died but that

> in a way, it doesn't matter to me which day my mother, already dysfunctional, happened technically to die. What I really want to know, and never can, is a different date, the date she was lost to the world. A date to be guessed at like the date of a conception, the date some secret, hidden line of communication went dead, static overwhelming it. The end of thought, not the end of breath. The eclipsing end of the last of her memories, like a message on a scrap of paper gradually singed and then in a moment consumed by flame. The end of memory itself. The day she really died.[23]

This intense stigmatization has influenced the formation of policy on Alzheimer's disease, reinforcing the tendency to privilege biomedical research over caregiving. It has also profoundly affected both the way people with dementia have been regarded by others and the way people have experienced dementia. In short, since its emergence in the late nineteenth century, the line between the senile and the normal has been an important feature in the structure of knowledge, policy, and the entire experience of aging in America.

CHAPTER ONE

The Stereotype of Senility in Late-Nineteenth-Century America

In February 1905, William Osler, the most famous and idolized physician in the United States, if not the world, delivered what would soon become perhaps the most infamous assessment of aging in American history. Men above 40 years of age were comparatively useless, he claimed, while men over 60 were completely useless and, to the extent that they retained their positions, in fact caused much harm: "As it can be maintained that all the great advances have come from men under 40, so the history of the world shows that a very large proportion of the evils may be traced to the sexagenarians—nearly all the great mistakes politically and socially, all of the worst poems, most of the bad pictures, a majority of the bad novels, not a few of the bad sermons and speeches." Given this situation, Osler jokingly outlined a proposal, taken from Trollope's satire *The Fixed Period,* for mandatory retirement at the age of 60 and, following a year comfortably ensconced at a college created for the purpose, a peaceful death by chloroform. "The teacher's life should have three periods," Osler then argued more seriously. "Study until twenty-five, investigation until forty, profession until sixty, at which age I would have him retired on a double allowance. Whether Anthony Trollope's suggestion of a college and chloroform should be carried out or not, I have become a little dubious, as my own time is getting so short."[1]

The occasion of these remarks was Osler's farewell address to a capacity crowd of colleagues and alumni at the Johns Hopkins University, where for the previous fifteen years he had labored to elevate a fledgling medical school to one of the premiere institutions in the country, a mission in which he was successful. At age 56, Osler, and even more so his wife, felt that the pace at Hopkins was costing him his health. Osler was leaving Hopkins to become Regius Professor of Medicine—a post that, however prestigious, was essentially to be a sinecure following his cutting-edge work at Hopkins. On a trip to England to consider the offer, Osler asked a leading British physician, "Do you think I'm sufficiently senile to become Regius Professor at Oxford?"[2] In this context, his comments on aging at his farewell address were clearly intended as a joke to soften the blow of his departure to his colleagues and the school—a self-deprecating suggestion that, as he was clearly past his prime and nearly useless himself, Hopkins would do well to be rid of him. Although his colleagues at Hopkins could not accept the proposition that his departure was anything but a terrible loss to them, they took his remarks in that spirit, and the speech was warmly applauded and followed by heartfelt tributes and the awarding of an honorary degree.[3]

But reporters immediately picked up the explosive potential of his remarks, and the next day headlines around the country trumpeted the news that the great doctor Osler regarded men over 40 useless and advocated mandatory retirement and euthanasia for men over 60, creating a public controversy that simmered for months. A steady flow of letters and columns excoriating Osler and defending the virtues of older people appeared in newspapers and magazines. The word *Oslerize* entered the vernacular, appearing even as the title of a popular song, and he received mounds of abusive letters.[4] Although he was evidently surprised and irritated by the angry reaction to his speech, he weathered the storm with characteristic grace and humor, and although never forgotten by the press, the episode did not lastingly tarnish his reputation. In any event, Osler never recanted the substance of his speech. He responded to the controversy with public statements to the press and a letter to the editor in the *New York Times* asserting that he had never seriously suggested that men over 60 be chloroformed but nonetheless reasserting his belief that most of the productive work of the world was done by men under 40 and defending his call for mandatory retirement for men over 60.[5] Osler included the "Fixed Period" address with a brief preface in the second and subsequent edition of *Aequanimitas,* a collection of his popular essays. In this preface he explained again that the chloroform proposal had merely been a joke intended to relieve the sadness of his departure and that "to one who had all his life been devoted to old men, it was not a little distressing to be placarded in a world-wide way as their sworn enemy, and to every man over sixty whose spirit I may have thus unwittingly bruised, I tender my heartfelt regrets." Nonetheless, Osler concluded the preface by asserting that "the discussion which followed my remarks had not changed, but rather strengthened my belief that the real work of life is done before the fortieth year and that after the sixtieth year it would be best for the world and best for themselves if men rested from their labours."[6]

As William Graebner has noted in his history of retirement, and Michael Bliss in his more recent biography of Osler, this was not simply a case of misunderstanding an innocent, if ill-advised, joke. Osler's speech created a stir because it addressed quite clearly and seriously one of the most vexing issues of the day—mandatory retirement and the barriers that even middle-aged workers were beginning to face. Osler stated, perhaps more clearly and certainly more brutally, what many managers and policymakers had come to believe about older workers—and what those workers most feared.[7] Osler's speech was part of a discourse that began in the late nineteenth century and continued through the 1930s concerning the impact of industrialization on the physical and social aspects of aging. This discourse generated a new set of images and meanings for senile mental deterioration which have retained currency in American society and culture ever since.

Historians of old age generally agree that, though there has never been a "golden era" in American history in which the elderly were unequivocally revered, the second half of the nineteenth century saw an intensification of anxiety about the aging process and hostility toward the aged.[8] This can be seen in the changes that occurred in the meaning of the word *senility* itself. Through the early nineteenth century it was a neutral term, signifying the condition of being old. At the turn of the century, for example, Thomas Jefferson could write of eagerly looking forward to the day he could "exchange the roar and tumult of bulls and bears, for the prattle of my grand-children and senile rest." But by the end of the nineteenth century, the word had been appropriated by the medical profession to describe age-related deterioration, particularly of the mental faculties, and had thus acquired a pejorative meaning.[9]

As I describe in later chapters, this pejorative meaning and the stereotype of senility that accompanied it have remained a salient feature of American culture. In fact, subsequent discourses about many aspects of aging, including the emergence of Alzheimer's disease, can best be understood as reactions to the late-nineteenth-century stereotype of senility. In this chapter, I describe how these negative stereotypes of senility arose in medical and popular discourse and how they were connected to broad transformations in American society and culture.

FROM METAPHOR TO POSTMORTEM:
THE MEDICAL MODEL OF AGING AND SENILITY

Osler's ideas about the inevitable mental deterioration of old age were a long way from Benjamin Rush's relatively positive characterization of old age that I described in the introduction. The veneration of the aged that could be seen even in Rush's description of old people with dementia was largely swept aside in the nineteenth century. The patriarchal authority and hierarchical order within family and community in which the veneration of the aged was rooted were increasingly eroded by the material and ideological forces of an expanding market economy and the broad democratization of American society. As historian Thomas Cole puts it, "the ideal of veneration [of the aged] ill suited a society of individuals who were less and less willing to submerge their identities or constrain their activities within the communal confines of family, church, and town." At the same time, hierarchical order and reciprocal obligation were replaced by an emphasis on internal self-restraint and personal responsibility. Somewhat paradoxically, the rejection of traditional forms of external social authority over the individual in American culture produced an especially rigid code of inner self-control. According to Cole, "the power of this code derived from its ability to discipline the fierce desire for material wealth through careful rationalization of bodily impulses. Choosing the morally disciplined, autonomous individual as the ultimate unit of secular authority, the spokesmen of 'civilized' morality placed special emphasis on physical control and on structuring human energies to meet the demands of the marketplace."[10]

The erosion of patriarchal authority and hierarchical order also entailed a new vision of God and his relationship to humanity. Calvinist ideas of the absolute sovereignty of God and the preordained fate of all men and women made little sense to individuals who believed in shaping their own earthly and eternal futures, and by the 1820s the lofty, inscrutable God of Calvin had given way to the rational, benevolent God that characterized the Second Great Awakening. Evangelical ministers of the antebellum era increasingly stressed the agency and responsibility of the individual in turning to God and leading a virtuous life and the rewards that a just God would bestow—in this world as well as the next—to those who did so. In this view, suffering and loss in old age were no longer a meaningful illustration of humanity's dependence on God but evidence of moral failure. "The decay of age, as of death, is the sinner's accomplishment," a minister thus preached in the 1840s. "Every old man, therefore, presents in his body the testimony of nature to SIN and DEATH. Two dread realities."[11] The deterioration of the aging body came to signify precisely the loss of inner self-control and inability to rationalize the body which were most threatening to the moral order during the rapid and sometimes chaotic transformation of American society during the nineteenth century.

In these circumstances, it was not surprising that physicians—whose professional identity centered on their claim to be able to understand and influence bodily processes—displaced clergymen as the recognized authorities on old age. By midcentury, a medical model of aging had become the dominant element in much discussion of old age in America. This model was deeply pessimistic about the prospects for aging and asserted that the aged retained few, if any, productive or even functional roles in modern society.[12]

American medical science in general, and psychiatry and neurology in particular, lagged behind their counterparts in Europe throughout the nineteenth century. Outside of a relatively small elite of European-trained doctors, Americans were not able to undertake the sort of precise clinical and pathological studies that were being done in France in the mid-nineteenth century and in Germany toward the end of the century.[13] Throughout the period, most American physicians writing on senile dementia approached the topic from a highly metaphoric framework that applied ideas from physics (inertia and entropy) to physical and mental deterioration.[14] Aging in general and senile dementia in particular were typically described in terms of an endogenous vitalism. Every person began life with a finite quantity of vital energy. As this vital energy was depleted in old age, the body and mental powers began to deteriorate. Behavior and environment could ameliorate this process to some degree, for the way a person lived and the stresses he or she experienced could accelerate or retard the rate at which the body's vital energy was depleted. But given a finite amount of vitality, deterioration in old age was understood to be inevitable. As American physicians acquired a more elaborate knowledge of pathology, this vitalistic model was replaced by more mechanistic theories of tissue decay and evolutionary cell degeneration. Floridly metaphoric language was replaced by the more concrete terminology of postmortem dissection and microscopic description. But the basic medical approach to old age remained the same—aging brought inevitable deterioration of physical and mental abilities.

The metaphoric concept of aging as the depletion of a limited quantity of vitality remained dominant in American medicine into the first decade of the twentieth century, more than fifty years after French clinicians linked disease—and the degeneration of old age—to pathological structures discovered in autopsy.

Historian Carole Haber attributes the resilience of the fund-of-vitality metaphor among American physicians to their lack of training in clinical and pathological research and to the fact that the concept worked well enough to explain the ill and decrepit elderly with whom they came into contact in their practice, usually in an institutional setting such as public almshouses or, increasingly toward the end of the century, state mental hospitals. The French clinicians' ability to correlate sclerosis, fibrosis, and cellular degeneration with particular diseases was widely known among American physicians, but they simply regarded these pathological processes as the symptoms of the depletion of vitality in old age.[15]

According to this idea, each individual was endowed at birth with a supply of energy to be used for growth and activity. By adulthood, that supply of energy was diminished, and the individual could, at best, maintain a fragile equilibrium. In old age the fund was depleted, and the body began to deteriorate. This concept explained the visible transformation of the body with the passage of time. It also explained the susceptibility of the aged to disease, even when physicians did not agree on its nature and cause. Whether the physician believed disease was an entity that entered the body from without, an imbalance of the humors, or the product of unhealthy stimulation, its manifestation in old age could be explained by a predisposing debility in the aged individual. Lacking vital energy, the aged body simply could not maintain itself.[16] The implication of this model of aging for mental health was obvious. As British psychiatrist Charles Mercier explained in a textbook that went through several American editions, dementia was "the natural condition of man in his declining years" and was inevitable whenever death was the result of "the natural expiry of the forces of life, and not by violence or the quasi-violence of disease."[17]

While the concept of a depleted fund of vitality was a proxy for biological processes of deterioration that could not yet be directly observed, the entire emotional and social life history of the aging person had to be invoked to explain the variability in the age of onset of dementia. Mercier explained this by comparing a person's endowment of vital energy at birth to the inertia of a rolling ball. The greater force with which the ball is initially rolled, the farther it will go. "So with the living organism; all lives receive at conception an impetus which is to carry them forward to the end, but the impulse is not equally strong in all." Moreover, dementia was often hastened by disease, social factors, or personal habits. As Mercier put it, "owing to the exceptionally rough nature of the ground, the velocity of the ball is materially diminished at an early stage in its career, and . . . comes prematurely to rest."[18]

The belief that personal and social factors played a role in the onset of demen-

tia left room for some hope that it could be delayed, perhaps until the body succumbed to a fatal ailment of some other cause.[19] Some physicians even thought
that senile dementia was not inevitable if only the individual would take care to
avoid the habits of dissipation—excessive drink and sexual activity, an immoder-
ate diet, chronic over- or underwork—which squandered the fund of vitality. "The
normal brain ought to be the very last of the organs to fail in its functions," Ralph
Lyman Parsons argued in 1896. But in too many cases the mental faculties of the
aged fail with or before the physical. Sadly, "many of these failures might have
been prevented, or at least delayed if wise counsel had been obtained and followed," he concluded. A person's endowment of vitality "can never be increased,
although it can and often is diminished."[20] This emphasis on the responsibility of
the individual to avoid the habits that led to dementia was clearly related to the
moralization of health in the evangelical religion of the antebellum period. It was
a short leap from lecturing on hygiene to preaching on sin and redemption. This
link can be seen quite clearly in the writing of William Alcott, an evangelical
health reformer who argued that his regimen of Christian hygiene would allow
people to avoid deterioration and disease in old age. "Old age, whenever it is
wretched, is made so by sin. . . . If Methuselah suffered from what we call the
infirmities of age, it was his own fault. God, his creator never intended it."[21]

Still, according to most physicians, if the worst effects of senility could perhaps
be avoided or delayed through proper attention to diet and hygiene, progressive
decay was inevitable and had to be recognized. Thus Parsons argued that with the
onset of "the physical infirmities and changes incident to old age"—which included loss of muscular strength, wrinkling of the skin, trembling of the hands,
emaciation, diminished tone and resiliency of the vascular system, dulling of the
senses, especially of the hearing, and much more—there also came important
mental changes such as "irritability, imperiousness, excitability or diminution of
normal emotional responses, loss of memory, diminished attention-span, diminished power of abstract thought, and fickleness or perversity of disposition." This
impressive list of calamities was normal; these were the changes that were to be
expected with old age and which constituted the "evidence of weakness that
forebodes impending danger" which the physician and his patient must heed.[22]
Once the worst effects of senility had appeared, there was little hope. As another
physician put it in discussing arteriosclerosis and senile dementia, "where so
little can be done to restore such cases to health, no extended discussion of
treatment seems to be called for."[23]

The pessimistic attitude of American physicians toward old age in general, and
senile dementia in particular, was fueled by the pathological findings of the great

French clinical schools, which were becoming better known in the United States. Increasingly in the last quarter of the nineteenth century, American physicians used the term *senile* to differentiate between diseases in the young, which were often curable, and diseases in the aged, which inevitably resulted in increased debility and eventual death. Even an illness that affected all age groups and which was usually considered self-limiting, such as bronchitis, became "senile bronchitis" when it afflicted the elderly—to indicate the more negative course and outcome that could be expected because of the deterioration of vascular tissue and mucous membranes which accompanied normal aging. The word *senile* also came to be associated with mental deterioration in particular. The brain, like the rest of the body, underwent inevitable deterioration with age, and the mental impairments commonly observed in the aged were the result. Although this usage of *senile* ostensibly added precision to medicine, allowing physicians to adjust their expectations and therapeutic strategies appropriately, the word became in practice a vague, elastic term that tended to pathologize aging itself.[24] Few psychiatrists would have disagreed with W. Bevan Lewis that "were we to attempt to define the boundary betwixt the physiological and the pathological forms of senility, between the ordinary second childishness of old age, and the dementia resulting from the senile atrophy of disease, we should find the task a difficult if not impossible one."[25]

As physicians in the United States became better trained in the techniques of clinical and pathological research, metaphoric language about the depletion of a "fund of vitality" was displaced by the language of postmortem examination of tissue. Descriptions like Mercier's analogy between dementia and a rolling ball became rare. Increasingly typical was the sort of description found in a 1902 article by William L. Russell, psychiatrist and superintendent at the Willard State Mental Hospital in New York State. "Sclerotic changes in the walls of the blood vessels" impair the "nutrition and vigor of the parts supplied by them," resulting in "fatty, pigmentary and calcerous degenerations" that further decrease the efficiency of circulation. "Atrophic and degenerative changes are especially conspicuous in the brain, the chemical composition of which is also modified, the fatty materials suffering diminution and the phosphorous and water being increased." Naturally, "alterations in function" were the inevitable consequence of these degenerative senile changes, and "the characteristic of these is retardation."[26]

By 1910, the medical literature on senile dementia had come to focus almost entirely on brain pathology. Nearly all the articles appearing between 1910 and 1930 in the *American Journal of Insanity* and the *Archives of Neurology and Psychiatry*, the two leading professional journals of psychological medicine,[27] concerned the pathological structures of senile dementia, cerebral arteriosclerosis and the recently named Alzheimer's disease.[28] (Chapter 2 tells the story of the emergence of Alzheimer's as a disease entity.) Thus, as metaphoric language disappeared from the medical literature, so too did prophylactic advice and any sense that changes in behavior could modify the onset or course of dementia. Indeed, the personal details of the patient's life ceased to have much relevance with the new emphasis on pathology in explaining dementia. In a typical case description in these articles, a cursory paragraph might be devoted to the patient's life history, followed by a lengthier description of the clinical picture and then several pages on the pathological findings in autopsy.[29] As psychiatrist E. E. Southard wrote in 1910, these physicians, most of them psychiatrists working in the state mental hospitals, found "little convincing evidence that social factors play much part in . . . these cases. Whatever the cause of brain atrophy, it seems that they cannot be social."[30]

However, if the medical literature on dementia was becoming more concrete in its description of brain pathology, it was no closer to being objective about old age. While psychiatrists of this period were writing about senile dementia in language that was much more similar to that of what we have come to speak of as modern scientific medicine, the profession as a whole had an increasingly negative attitude toward aged patients. In addition to the problematic position of old age in American culture, there were more concrete reasons for this attitude: the older people psychiatrists saw in their practice in the state mental hospital system were typically frail and suffering from dementia. Psychiatrists saw these patients as an increasing threat to the therapeutic integrity of the state mental hospitals, threatening to transform them into warehouses for the incurable. This in turn threatened the authority and prestige of the psychiatric profession itself, which through the mid-twentieth century were based on the prominence and importance of these institutions as vehicles for the successful treatment of mental disorders.[31] These institutions were the basis of psychiatric therapeutics during this period, providing an environment of moral stability thought to be curative for those experiencing an acute episode of mental illness, and through the mid-nineteenth century they reported very high recovery rates.[32] Whether they thought senile dementia resulted from the depletion of irreplaceable vital energy or from irreparable arteriosclerotic deterioration, nearly all psychiatrists agreed that it was by definition incurable and hence that patients who had senile dementia should be institutionalized elsewhere. But public policy changes that I describe in more detail in chapter 2 dictated the opposite, and elderly patients became an increasingly large segment of the patient population in state mental hospitals. As a result, aged patients were, by the end of the century, the bête noire of the psychiatric profession. Psychiatry's particular problem with the elderly was well summarized by F. H. Stephenson in 1901, when he complained that "these old people are often very troublesome and have to be committed to hospitals for the insane." The protection and routine of the hospital often had a beneficial effect on them, but "many realize their confinement, and considering it unjust, they make a great disturbance for all about them and seem the most trying of all lunatics."[33] Thus, if psychiatrists and other physicians were groping toward a better understanding of the brain and the processes of deterioration that resulted in dementia, they were also reiterating the negative stereotype of the elderly that emerged in American culture during the antebellum period.

DYING AT THE TOP FIRST:
MODERN REPRESENTATIONS OF SENILITY

There is no better place to begin exploring the contours of that stereotype than with George Miller Beard, the New York neurologist who popularized the diagnosis of neurasthenia—an ailment that many historians have seen as emblematic of late-nineteenth-century American anxiety about the pace of modern industrial society.[34] Beard himself is often seen in a similar light. Historian Charles Rosenberg argues that Beard should be understood not as a creative or original thinker but as a writer uniquely reflective of his period. "His medical writings are a mosaic of the intellectual commonplaces of his time; it was the familiarity rather than the novelty of Beard's theories which made them so easily and rapidly accepted. . . . His pages record with ingenuous fidelity the intellectual temper of post-bellum America."[35] Historians of aging have seen him as one of the principal architects of the scientific legitimization of the denigration of old age.[36]

In Beard's *Legal Responsibility in Old Age*, first delivered as a lecture, published as a single volume in 1874, and then incorporated in perhaps his most famous work, *American Nervousness*, published in 1881, we can see the modern stereotype of senility in some of its most characteristic forms. In this work, Beard described the average effect of age on the mental faculties—what he tellingly called "the law of the relation of age to work," which revealed that aging brought about a catastrophic decline in productivity. Beard derived this law by correlating great works with the age of their creators—studying "in detail the biographies of distinguished men and women of every age" and tabulating the ages at which their greatest work appeared.[37]

Beard found that 70 percent of the work of the world is done before age 45, and

80 percent before age 50, and that "on the average, the last *twenty* years of a genius's life were unproductive." Beard concluded that, like the rest of the body, the brain and the mental faculties that depended on it were subject to deterioration and in fact would normally do so more rapidly. "Men die as trees die, slowly, and frequently at the top first," he argued. "As the moral and reasoning faculties are the highest, most complex and most delicate development of human nature, they are the first to show signs of cerebral disease; when they begin to decay in advanced life we are generally safe in predicting that, if neglected, other faculties will sooner or later be impaired. When conscience is gone the constitution may soon follow."[38] For Beard, senile mental deterioration could hold no consolations; it signified only the start of the gradual, vegetative process of death—the slow, topdown withering away of a life.

The dissolution of the intellectual and moral faculties was not necessarily a simple process, however. "Very few men decline in all the moral faculties," Beard observed. "One becomes peevish, another avaricious, another misanthropic, another mean and tyrannical, another exacting and querulous, another sensual, another cold and cruelly conservative, another vain and ambitious, and others simply lose their moral enthusiasm, or their moral courage, or their capacity of resisting temptation and enduring disappointment." Beard thus represented the elderly in a number of characteristic guises. In his description of editor and politico Horace Greeley, for example, Beard represented the elderly as miser: "He [Greeley], who all of his life had cared nothing for money, in his old age, when disease had invaded his brain, cared for little else; and all this time, or most of it, he was capable of conversing intelligently, and of writing very good editorials." Interestingly, although Beard saw Greeley's concern for money as a pathological sign, elsewhere he remarked in a formulation strikingly similar to Rush's "that the faculty for business," which surely must involve caring intensely for money, "sometimes lingers in the brain long after the other faculties have fled forever. An old man may know well enough how to look out for his money when on almost any other subject he is perfectly childish." In another of Beard's descriptions, that of a clergyman who between the ages of 70 and 80 "displayed a morbid sexual appetite, and went to gross excesses," the representation was that of a dirty old man.[39]

Whatever the particular form, Beard represented senile old people as hopelessly out of step with the times. Failing to fulfill the complex intellectual and moral tasks required of the individual in modern society, such people were ultimately an obstruction to progress. Beard thought that the great biologist Louis Agassiz, who died shortly before *Legal Responsibilities* was published, was a perfect example. "The intemperate manner of his opposition to the theory of evolution by which he was so rapidly winning favor among the thoughtless and ignorant, and so rapidly losing favor among the conscientious and scholarly, may find its partial, if not complete explanation in the exhausted condition of his brain."[40]

In pointing to Agassiz's unwarranted influence on the "thoughtless and ignorant," Beard represented old people as dangerous. Given the grim reality of the law of the relation of age to work, society could no longer afford to indulge a reflexive reverence for the aged. Beard thought that reverence for the past accomplishments of age, for "the really great and good thoughts it gave to the world," was appropriate. But the majority, "ignorant of the fact that reason, like the muscular and osseous system, undergoes various changes and degeneration in advancing years," revered the aged blindly and disproportionately and were credulously persuaded by the opinions of the aged on the merits of past accomplishments or, what was worse, on the merits of age alone. In the case of Agassiz, "conscientious and scholarly" men, men possessed of a "trained mind" in the prime of its powers, would revere Agassiz for "the *original* work that he did before his *fortieth* year." But they would not be influenced by his late opinions, which were clearly the product of the mental deterioration of aging, a sign of what was "well known" to Agassiz's friends—that he actually "began to die" long before his death in 1873. Beard's representations of old men as dangerous extended also to practical affairs of public and private life, where they were perhaps even more disruptive. "Corruption in political and business life, and breaches of trust, are very common among the old, as every morning newspaper bears witness," Beard asserted, and offenses "that depend on the sexual passion are not infrequent among the aged; for it is a fact of interest that in the decline of life we sometimes return to the vices of youth."[41]

These stereotypical representations of senility were also clearly evident in the work of the American physician Ignatz Leo Nascher, who coined the word *geriatrics* and is widely regarded today as the founder of the field.[42] Nowhere was the stereotype of senility more apparent than when Nascher tried to make a case for the formal recognition of geriatrics as a specialty analogous to pediatrics, as in his 1914 textbook *Geriatrics*. The logical rationale for such a specialty was clear: the normal individual in senility presented marked deviations from the normal individual in maturity; failure to take these age-associated changes into account would result in the misdiagnosis and mistreatment of disease. But Nascher understood that the real barrier to the development of a specialty was the "natural" abhorrence of old age. "The appearance of the senile individual is repellent both to the esthetic sense and to the sense of independence, that sense or mental attitude that the human race holds toward the self-reliant and self-dependent," Nascher observed. This was, of course, not a barrier to pediatrics, where "the dependence of the child arouses sympathy."[43]

Nascher thought this abhorrence quite understandable, and nowhere in his text is there anything to suggest that he did not share it. "We realize that for all practical purposes the lives of the aged are useless, that they are often a burden to themselves, their family and the community at large," he wrote in the preface to *Geriatrics*. "Their appearance is generally unesthetic, their actions objectionable, their very existence often an incubus to those who in a spirit of humanity or duty take upon themselves the care of the aged." Among physicians Nascher acknowledged "a natural reluctance to exert oneself for those who are economically worthless and must remain so, or to strive against the inevitable, though there be the possibility of momentary success, or to devote time and effort in so unfruitful a field when both can be used to greater material advantage in other fields of medicine." But Nascher thought the physician would set this "natural reluctance" aside out of scientific interest and because "as a humanitarian it is his duty to prolong life as long as there is life and to relieve distress wherever he may find it."[44]

Although Nascher used the word *senility* to describe the age-related deterioration of all functions of the body and was concerned with describing and differentiating between this normal deterioration and disease (the physiological and the pathological) in every organ of the body, he, like Beard, gave special emphasis to the deterioration of the brain and mental faculties because it was often the first sign of senile deterioration. "Lessened interest in the events of the day, a tendency to sleep after some mental work, greater difficulty in getting ideas or some particular word to express ideas, forgetfulness, all point to senile changes in the brain."[45]

Nascher's representations of the senile, like Beard's, took characteristic and familiar forms. Nascher frequently represented the senile old man as garrulously self-centered and simpering. "The old man's talk begins with 'I' and ends with 'me,'" Nascher asserted. "When the mind becomes impaired he neglects his person in every direction until he becomes obnoxious to those around him. . . . He demands constant attention and complains of the slightest neglect. The firm insistence upon hygienic measures for his benefit and welfare, which necessarily impose some exertion on his part, is resented as a hardship and creates a dislike of those who are most interested in his welfare."[46]

Nascher was particularly concerned with sexuality in the aged, and the representation of the dirty old man appears frequently in his writing. Nascher devoted a short chapter of *Geriatrics* to "sexual perversions," which he found to be "rather common in the aged and due to either diminished power with undiminished desire or to recrudescence of desire after power and desire had previously disappeared." Nascher did not believe that senile sexual perversion indicated "depravity or inverted sexuality," as it did in the young. Rather, it was simply a reflection of "weakened mentality, diminished control over the emotions and some circumstances producing intense emotional excitement." In most cases, Nascher noted, the individual—always figured male—is "unable to perform the sexual act and adopts extraordinary measures to gratify his desire."[47]

As these examples illustrate, senility was characteristically represented as a problem for men. But senility was also represented in terms of class as well as gender. Figures described as representative specimens of senility so clearly reflect middle-class occupations and living circumstances that senility seemed a decidedly middle-class malady. In Beard's case, this flowed directly out of his theory of nervous exhaustion. Senile deterioration, like neurasthenia, was the product of the overuse of the mental faculties among middle-class "brain workers." Lower economic classes are represented briefly in Nascher's chapter on senile dementia and suggest an explanation for their absence as representative cases. Nascher argued that brain cells, like all cells, could deteriorate through underuse or overuse. "Persons of low mentality, such as the uneducated peasants," clearly represented the former case and found it "almost impossible to learn anything new after their thirtieth or thirty-fifth year." In a phrase that clearly marks these people as part of the flood of immigrants that so troubled middle-class America at the turn of the century, Nascher, who was born in Vienna and brought to the United States as an infant, noted that "they may retain the memory of events that have made a powerful impression upon them but they cannot learn a new language, though spending years among those who speak that language."[48] Thus, one supposes, these "persons of low mentality" experienced a premature mental deterioration that precluded their figuring as representative cases of senility.

Ultimately, the impressive variety of moral and intellectual failures that Beard and Nascher used to represent senility can be read as a catalog of nineteenth-century middle-class male anxieties. The dominant gender ideology in the Victorian era figured men as aggressively competitive and bold; the era was characterized by stunning economic growth and the acquisition of fabulous fortunes by a few and grinding poverty for many; there was boundless optimism in progress yet gnawing dread of social disorder. The moral failings and calamities that Beard ascribed to the the old threatened to undo a man at every stage of his life in this highly individualistic and competitive society. Projecting these failings onto a figure—the aged man—visibly marked as decadent and diseased helped restore a sense of moral stability. Whatever particular guise that figure might take, he was always and above all the very picture of the ruined man. But if the doddering old man, afflicted with cerebral degeneration of one sort or another, would inevitably unravel under the conflicting pressures of the modern world, the young man, in his intellectual and moral prime, might well expect to flourish.

"THE PASSING OF THE OLD LADY": THE STEREOTYPE OF THE PICTURESQUE GRANDMA AS A CALL TO DOMESTICITY

Given that representations of senility bore this burden of middle-class male anxiety about measuring up to the complex, grinding tasks a man faced in modern industrial society, it is not surprising that they were so strongly figured male. No concrete representation of a woman appears anywhere in Beard's text, and the only explicit reference is an abstract observation that "in woman, the procreative function ceases between forty and fifty, just the time when the physical and mental powers begin to decline, as though nature had foreseen this law and provided that the world should not be peopled by those whose powers had fallen from their maximum."[49] The absence of any representation of women save in reference to procreation clearly reflects the gendered social order. As Beard was concerned with the effect of senile mental deterioration on work and legal responsibility— the public sphere—old women had no place in his representation of senility.

But the figure of the old woman did make a fleeting appearance in Osler's "Fixed Period" address of 1905, which so strongly echoed Beard. After sketching out his ideas on the uselessness of older men and the "remedy" of chloroform, Osler noted, "for the benefit of the public," that "with a woman I would advise an entirely different plan, since, after sixty her influence on her sex may be most helpful, particularly if aided by those charming accessories, a cap and a fichu."[50] Osler did not need to elaborate on the salutary influence this charming old woman would exert, for in listing her props—the cap and fichu—he summoned up one of the most powerful images of domesticity in that era, a tug on the heartstrings of young women that would help to hold the home together.

This image of the frail, bespectacled old grandmother, wrapped decorously in her shawl, sitting by the fireside, absorbed in some old-fashioned devotion to domesticity such as knitting or embroidery, was a call to defend domestic virtues that were supposedly threatened by the aggressive, competitive individualism of the public world. Thus while the aged man's purported inability to change with the times was a sign of pathology that marked him as useless, the same inability in the aged woman was a sign of her virtue and the continued contribution she could make. Thus Lucy Martin Donnelly, writing in 1899 in the *Atlantic* of her lifelong fondness for old ladies, argued that:

> it may be that the old gentleman is unamiable; that, his days of strenuousness fairly over, he becomes crabbed, a lover of snuff, and unpoetical. But the old lady is a creature of another quality. The refinements of age only enhance the femininity of her charm; to her, whimsicalities, delicate occupations, the fine lines that etch themselves expressively across her brow and about her mouth, are all vastly becoming. With what ineffable grace, moreover, she pronounces certain words in the elegant fashion of an age ago! How softly the old India shawls she wears fall about her shoulders! What strange, unlikely stories she tells of the beginning of the century![51]

As the emergence of the New Woman challenged the Victorian ideal of domesticity during this period, the stereotype of the Old Lady became a point of contention. An exchange in 1907 between anonymous authors in the *Atlantic*'s "Contributor's Club" illustrates this. (Though neither author explicitly revealed gender, the easy familiarity with which they discussed domestic life suggests that they were women.) Lamenting the "passing of the Old Lady," the first contributor held out the hope that after the "modern substitute" for the Old Lady died from "over-athleticism, or exposing herself to a draft in a low-necked gown" in her frantic pursuit of youth, the "dear old lady of a past era will awake, pick up the dropped stitches of her knitting, rub her spectacles, and resume her interrupted sway." At stake, this author thought, were the manners and morals of the young. The elderly had to be convinced that "in defying age they are corrupting youth." She concluded that "the old lady must be born again . . . [for] the New Woman can never grow into the Old Lady."[52] In reply, the second author began by challenging this image of the Old Lady. "Thank heaven there were and are such old ladies— enough of them to fix the type. . . . But there were also restless old ladies, critical and carping old ladies, interfering old ladies—old ladies who cultivated in their daughters and daughters-in-law those graces of forbearance and unselfishness in which they were themselves so conspicuously lacking." Given the Old Lady's potential to disrupt the household, it was "one of the great blessings of modern life that in retaining their physical vigor longer than of old, women also retain their own pursuits." The author was unwilling to completely destroy the image of the traditional Old Lady: she admits, not without sarcasm, that the "new" old ladies displayed poor taste in dressing as though they were young and were thus "less picturesque than the capped and kerchiefed ornament of the domestic fire-side." But she ultimately represented the figure of the traditional Old Lady, however delightfully quaint she may be, as a barrier to the progress that the New Woman represents, for, as the first author had noted, the New Woman would never grow into the traditional Old Lady. "Make a houseplant of the young girl," she argued, "dispense with modern ideas of hygiene, and you may get back your old lady of the fireside."[53]

Not surprisingly, when physicians described senility in women, their representations reflected the stereotypes generated by the discourse on the disappearing Old Lady. Echoes of all three stereotypes are clearly seen, for example, in Nascher's work. In an abstract differentiation of the appearance of senility in women, Nascher clearly represented the senile woman as the traditional Old Lady:

> The obvious manifestations of senility appear later in the female, for the reason that she makes an effort to remain attractive, the psychic factor involved in the production of the senile slouch in the male being overcome by her vanity, there is absent the senile kyphosis [curvature of the spine], the marked waste of the facial muscles, and often the wrinkles generally seen in the male. Women being more impressionable than men, they are more amenable to religious teachings, they become more readily resigned to the inevitable through their faith and hope of eternal life hereafter, and being more cheerful they do not present the disagreeable, gloomy appearance of aged men. This as well as their sex bring to them the sympathy denied to men.[54]

In describing a particular case, however, Nascher was as likely to represent the senile woman as the critical, carping Old Lady. Thus a woman who until 60 had been self-supporting exhibited a "complete change of personality" soon after moving in with a married daughter. She became "peevish and complaining," was "susceptible to flattery but could not stand fault-finding, and if one of her children did attempt to correct her" would remove that son or daughter from her will.[55] Yet adopting the fashion, interests, and pursuits of youth might also be figured in Nascher's representations as a sign of senility in women. Thus when an 82-year-old woman whose association until she was 80 had been "with persons her own age and station" and who had "dressed and acted accordingly" began to pursue the interests of the "New Woman" of the younger generation—"including woman suffrage, riding astride, mother teachers, female invasion of masculine occupations, and all that"—Nascher marked this as a sign of senile mental impairment. Although this woman's memory was seriously impaired, her conversation was "rational and coherent, although inappropriate for a woman of her age; and to the casual observer she appears giddy and silly."[56]

Turn-of-the-century representations of the senile woman, like those of the her male counterpart, thus clearly bore the burden of anxiety about the stability of women's identity as Victorian gender norms showed signs of unraveling.

"A CASTE APART": RECOGNIZING SENILITY IN THE OTHER AND IN THE SELF

For all the clarity with which they stereotypically represented senility, medical experts stressed the difficulty of recognizing it. "Responsibility has passed into irresponsibility. Clearly responsible ten years ago, clearly irresponsible now, where was the dividing line?" Beard asked of a typically complicated case. "When a man has passed into driveling idiocy or helpless imbecility Macaulay's school boy can make the diagnosis," he noted. "But to tell just where idiocy or imbecility began may defy the ablest physiologist or pathologist."[57] Driven by the danger they saw in senility and by their own anxiety, medical experts subjected the body and behavior of the elderly to increasingly intense scrutiny in an effort to discern the line that divided them from the rest of humanity.

Nowhere is this tendency better illustrated than in a 1915 article in which Nascher painstakingly scrutinized a man he had known socially for more than twenty years in order to discuss how the physician might recognize the early signs of senility. As Nascher began to note "the little peculiarities that pointed to mental impairment," every aspect of this aging man's life came under question

At the age of 50, the man had been successful in business, prominent in fraternal organizations (evidently the context in which Nascher knew him), a forceful speaker, systematic reader, genial, and popular. He made an excellent appearance and was vain of "a fine head of hair and long well-kept whiskers." At about age 55, he showed signs of anxiety about his age, using hair dye and aphrodisiacs. When the man was 60, Nascher began to note signs of deterioration. He began to lose interest in the affairs of the day but displayed "great interest in lodge matters where his strict rulings in cases where common sense and sentiment dictated a deviation from parliamentary practice made him unpopular." His memory began to fail, forcing him frequently to "look up authorities and decisions which he formerly could quote offhand."

Within five years, Nascher thought that the mental impairment was obvious to the close observer. Although the man continued to use hair dyes, "his hair and whiskers were getting thin and he frequently ran his fingers through the hair, pleased when no hairs came out and dejected when he found a hair or two adhering to them." He now took little interest even in lodge matters and was frequently depressed. He grew indifferent to his family, became absorbed in his business—which was operating at a loss because of his obstinate adherence to antiquated methods and machinery—and grew increasingly anxious about his age. He recognized his own weakness in memory and reasoning and tried to cover these up by avoiding conversation. He "became a slave to a routine mode of life," going back and forth from his home to his shop "like an automaton." When, on several occasions, Nascher met him on the way to his shop or home, the man "passed by staring ahead or, if I stood in his way, he passed around, seemingly unconscious that there was an animate obstruction in his path." When Nascher spoke to him, "he appeared startled."

Nascher concluded this case history with two events that occurred around the man's seventieth birthday. First, the combination of physical debility and a fire in the building where his shop was located forced him to retire from business. After the fire he would absentmindedly walk to the burned building and attempt to open the door with his keys. "It would be necessary to show him the burnt interior, before he could realize that he could not enter." The second event was the man's seventieth birthday party, for which he had prepared a speech, the "manuscript of which showed errors in spelling, omission of words, introduction of foreign matter, numbers and scrawls." He tried to memorize the speech while walking through the streets and in his rooms. At the celebration, as congratulatory speeches were being made, he kept mumbling to himself, rehearsing the speech, oblivious to his surroundings. "When he was prompted to arise he repeated a few of his lines, then dimly realizing he could not do what was expected of him he broke down and wept, only to be fast asleep a few minutes later."

Following these events, the mental decay proceeded rapidly. He soon forgot the names of his wife and children. "For several months before he died he kept up an unintelligible jabber, did not recognize anyone . . . and toward the end of his existence was purely vegetative." Nascher summarized the case by allowing that "the excitement connected with the burning of his place of business, as also the birthday celebration, probably hastened the final breakdown, but in its entirety, this is a perfect case of progressive senile mental degeneration, from strong mentality to complete decay."[58]

Reading this case today gives us a sense of how deeply the representational line that defined senility cut into a life. Not only did it call into question every aspect of life in the present, but it also extended the scrutiny back into the past. When the man was 65 and unable to hide his mental impairment from a close observer, Nascher had already been noting "little peculiarities" for five years and, in telling the case at least, extended this history back ten more years still to the age of 50. What was worse, though Nascher did not explicitly mention it, the line his observations traced extended into this man's future as well. When the first signs of deterioration were present, Nascher must have foreseen the likelihood of utter dementia, a vegetative existence, and death. Ultimately, we also sense how important it was for Nascher to find that line, to sift through the charred remains of a ruined life, to poke around the burned interior to assure himself that these signs of failure and dissolution were but the clinical signs of a distant pathology.

Of course, this pathology did not seem distant to those who themselves were under scrutiny. We cannot know precisely what Nascher's ruined businessman felt at different points, but, if Nascher's observations are accurate, shame, humiliation, and despair are apparent in his failed attempts to hide his aging, cover up his failing memory, and deliver a coherent speech on the occasion of his birthday. Surely this ruined businessman was as aware as Nascher, and probably more aware, that his failure to perform these social tasks put him over the line that defined senility.

Sander Gilman has argued that the internalization of stereotypes of the diseased Other is perhaps the most salient feature in the experience of disease. The individuals onto whom cultural anxieties about the dissolution of self are projected—in this case older people who are labeled senile—share those anxieties and so are put in a dilemma. Sharing in the same desire to domesticate anxiety about the dissolution of the self, the Other must continue to identify with society and so cannot reject social categories of Otherness. Yet he or she must somehow reconcile his or her own sense of self with the image of the Other, projected onto a group with which he or she has been identified. Paradoxically, those labeled senile must accept society's sense of their difference in order to re-create a sense of oneness with the world. Typical strategies for resolving this dilemma involve denying membership in the disease category, and thus legitimating the basic categorization, or affirming membership but identifying subgroups onto which the negative associations can be projected. "The image of the diseased makes those so categorized need to prove their basic healthiness," Gilman argues.[59]

The dilemma created by the stereotype of senility is clearly evident in the popular discourse surrounding old age during this period, in which mass-circulation magazines published hundreds of articles advising their readership on ways to avoid old age and senility.[60] The appeal of these articles clearly lay in the hope they offered that the ravages of aging might be escaped, but in offering this hope they actually reinforced the stereotype. By force of will, careful attention to hygiene and exercise, seeking out the companionship of the young, and many other stratagems, the white middle-class men and women who were confronting the age line could differentiate themselves from the decrepit, senile Other.[61]

This struggle to reconcile images of self and Other could be clearly seen in an article by "poet and philosopher" Walt Mason, a widely syndicated journalist and humorist in the first two decades of the twentieth century.[62] The magazine's editor introduced the piece, entitled "I Refuse to Grow Old," by noting that "many a man gets shoved out of his job when he is forty. People say: 'He's too old.' He may be all right physically—but an old fogy, mentally. Mason shows us how to beat the game." Mason's secret was to battle the habits and behaviors that indicated "senile decay." Facing the abyss of old age shortly after his fiftieth birthday, Mason determined that "age is a state of mind more than anything else." He began to pay greater attention to his appearance, especially to his clothes. "The importance of this scheme to an aging man cannot be overestimated," Mason argued. "It is the antidote to that general sagging down which makes old men bores. When one is well dressed he naturally strives to have everything harmonious; so he walks with vigor, and avoids the stoop and shamble, and really gives a good imitation of a young man, until someone looks in his mouth and sees that the cusps are all worn out of his teeth, and he can no longer eat corn off the cob." Still, Mason concluded, dressing well "does more than anything else to make a man feel young, and feeling young is being young."[63] The tension in this passage between Mason's desire to appear young and his inevitable failure to completely hide the deterioration of aging is typical of the entire text. The humorous tone served to distance author and reader from the deeply disturbing position of the aging man, who was subject to the strict scrutiny of a society intent to discern in him the signs of senility. The only hope the aging man had of avoiding senility was, through an even more intensive self-scrutiny, to discover and disguise its signs before they were noticed by society.

Mason's self-scrutiny was most explicit in his effort to avoid old fogyism in thought and conversation. He began by noting that he had studied "a lot of old men" that he knew, "old men who meant well . . . but who are avoided because they are tiresome." Their offense was a tendency to cling to the past. "They were facing backward, every man jack of them," Mason asserted, and it was "this unfortunate weakness for looking backward that lends some justification to Osler's theory . . . that the old boys should fall off the earth." Old men tended to believe that things were better when they were young and to rail against innovations of any kind. Mason found that he was beginning to think that way himself, "arguing that the world was a noble place when I was young; the men had nobler ambitions, and the women made better pies." But when he began talking this way in the grocery store, he noted that "people would drift out, and the grocer would hint that a long walk would be an excellent thing for a man of my sedentary habits." He resolved to squelch this tendency to look backward and "shifted right around and began explaining to the boys that this world, as it now exists, is an excellent resort, and that it's bound to improve right along." But to succeed at this would require constant self-scrutiny. "Eternal vigilance is the price of evergreen youth. One must forever be on his guard, or he will fall into some of the habits which indicate the dotard."[64]

As we have seen, representations of old women in this period typically depicted them as the embodiment of a feminine commitment to hearth and home. But if aging women did not have to struggle directly, as men did, with the belief that they were worn out and obsolete, representations of the "picturesque grandma" were at least as burdensome. Where middle-class men had to contend primarily with a pathological view of aging that suggested that they could no longer participate in the productive world of work, middle-class women had to contend primarily with a sentimental view of aging that held that they should more than ever restrict their interests and activities to the emotional world of the family.

An anonymous "elderly woman" described her struggle against this stereotype in a series of popular magazine articles published in the first decade of the century.[65] In one article, she reacted to the typical complaint of a young woman that "there are no more real grandmothers left in the world!" with irritation at the conventions the young pressed on the old. "Each generation permits a different type of young girl, but the older woman must not change. Her outline is fixed and immovable," the author lamented. "The elderly woman must be always there, waiting, waiting, with a smiling face through long, quiet, empty hours, for her grandchildren to come home." This was all the more galling because it capped a lifetime of conforming to the ideals of others. The young woman must take care "to avoid even the appearance of evil," and later, as a mother, she must follow conventions to set an example for her children. "But when a woman grows old, when she has climbed the ladder of years beyond the point where scandal could touch her, one would think that she might lay aside minor conventions of life, that at last she might do what she pleased, only limited by her own failing strength." But such was not the case, for when she arrived at old age, a woman found waiting for her yet another ideal of what sort of person she should be—"a graceful, shadowy person, sitting, her feet on a hassock—like Whistler's mother—someone who has none of the impulses of youth" that "in a grandmother, the younger generation find so disconcerting." The implications of this image, the author argues, were that the old woman's body had "so lost its resiliency that the wholesome desire for action has passed, that one's own life and actions have ceased to have an interest for one, and that instead of having to snatch time and make time to play with one's grandchildren one has nothing to do but wait—nothing in the world to do but 'be there.' "[66]

The author pitied women who lived up to this image of the ideal grandmother and resolved to resist it herself. "It is too great a price to pay for conforming to an ideal whose greatest value, after all, lies in a certain picturesqueness." She cheered for those older women who resisted in whatever small ways they could. For example, she described an elderly friend of hers who, having always loved to wear pink, found it a real cross to have grown too old to wear it any longer. The author asked her why she did not simply wear it anyway, though she knew the answer, since she herself "would not have the courage to blossom out in so much as one daring pink ribbon." To her surprise, her friend replied that she, in fact, did. "I looked at her simple black gown. 'Oh, not on the *outside*! but' said she, lowering her voice, 'I always run in plenty of pink ribbons in my things, and I have pink ribbon garters!' . . . Only a woman who has been cut off by an arbitrary custom from many of the pretty gay things of life will understand what a comfort those pink ribbon garters were to her."[67]

But the secret comfort of hidden pink ribbons could hardly resolve her struggles with the ideal image of the old woman with which she struggled. If older women did not face the "age deadline" in the labor market, they had to struggle to retain autonomy even within their home. The author describes how she felt "a certain indignation" when thinking of how "I can visit my own attic and my own cellar only by stealth or with a devoted and tyrannical child of mine standing over me to see that I don't 'overdo.' For the motto of all devoted sons and daughters is: 'Nag mother to death, if necessary, but don't let her overdo.' "[68] Encounters with the young girls often made her feel inadequate, if not grotesque, as when, riding the trolley with a friend, two young girls sprang up to offer their seats to them, thus reminding them that "we belonged to the white-haired company who have won their right to a perpetual seat on the cars. And however welcome a seat may be, it is not so pleasant always to remember why it is our right."[69] Describing her young great-niece, the author noted with some pride that "she is pretty, as are her charming clothes; she looks one straight in the eye when she talks. . . . Though it is plain to be seen that none of the things which make a woman of one have touched her, she has a calm assurance of bearing that comes from perfect health." But whatever pride she feels in this young member of the family is tinged with a feeling of shame at her own deterioration. "Health, indeed, shines out of her; her vitality seems a force, and an almost overpowering one. In her presence I feel myself small and shrunken of body."[70] More generally, she felt intimidated by the young. "The strong new generation, eager and clamorous, is at your heels ready to take your place, anxious to perform your tasks. Already your children are altering the world that you know," she wrote. "One day you will wake up in a new world, an unhomelike place to which you must adjust yourself as a baby must adjust himself to his surroundings; but with the difference that every day the baby makes progress, whereas every day you will find the new conditions harder to understand—as I have, and as your mother has."[71] Overall, this author's account suggests that the stereotype of the "picturesque grandma" fostered a sense of captivity, humiliation, and loneliness for aging women.

Such themes were also evident in a notable scholarly book of the early twentieth century, G. Stanley Hall's *Senescence*—written when he was in his seventies, following his retirement as president of Clark University. In the preface to this work, Hall noted that once "we are thought of as old, whether because of mental or physical signs," there is no avoiding a sense of isolation. Many things reinforced the awareness of age. "Friends and perhaps even critics show that they take it into consideration. Shortcomings that date from earlier years are now ascribed to age. We feel, often falsely, that we are observed or even spied upon for signs of its approach, and we are constantly tempted to do or say things that show that it is not yet upon us." In all these ways, Hall argued, "isolation and the enhanced individuation characteristic of age separate us until in fact we feel more or less a caste apart."[72]

Unlike much of the popular discourse surrounding old age, Hall did not seek to deny age. Rather, he saw the very real Otherness that aging brought as the basis of an important but as yet unacknowledged and unfulfilled role for the elderly, for only the aged could really understand what old age meant. "There is a rapport between us oldsters, and we understand each other almost esoterically," Hall asserted. "We must accept and recognize this better knowledge of this stage of life as part of our present duty to the community." Self-scrutiny remained the only means of escaping the uselessness of senility, but rather than scrutinizing the self to disguise the signs of aging, Hall strove to create a category that transcended the negative image of the senile—the self-conscious elder who alone could understand what "ripe and normal age really is, means, can, should, and now must do, if our race is ever to achieve its true goal."[73] The bulk of the work is devoted to an unsentimental and at times clearly painful reading of the entire scientific, philosophical, and artistic literature on old age. Thomas Cole has shown the ways in which Hall anticipated, indeed prophetically demanded, the discipline of gerontology that emerged after World War II.[74] But, as Cole recognized, Hall's vision ultimately failed to transcend stereotypical representations of the elderly. Hall's creation of a subgroup of aged who, by virtue of their honesty and self-awareness, transcended the uselessness society attributed to old age simultaneously rein- forced the original negative categorization. By demonstrating his membership in this subgroup, Hall's self-aware elder could perhaps escape a useless old age. But, having internalized the stereotype, he or she could never escape the question itself.

Understanding the basic stereotype of senility is crucial to understanding the subsequent history of senile dementia and Alzheimer's disease. As Gilman ar- gues, the way the patient is perceived "structures the patient's treatment, the patient's status, the patient's self-understanding, as well as the patient's response to the complex interaction of social and biological forces that we call 'disease.' "[75] The image of senility that emerged around the turn of the century constituted the problem at which much research and policy in the twentieth century have been aimed. To be sure, the line anxiously drawn in the late nineteenth century to divide the demented from the rest of us has been a tenuous, shifting line— difficult at all times to trace. But it is a line that palpably remains with us, a line that must be drawn and redrawn—yet somehow never crossed. It is a line that has structured knowledge, policy, and the experience of every aging individual.

Beyond the Characteristic
Plaques and Tangles

Over the first half of the twentieth century, American psychiatry became increasingly interested in the problem of senile dementia. At least initially, psychiatrists did not find it inherently interesting as a medical problem. As the medical literature described in the previous chapter suggests, senile mental deterioration did seem rather easy to explain away as an inevitable part of aging for which there appeared no prospect for meaningful intervention. Nor were psychiatrists particularly interested in helping older people. All evidence suggests that they were as anxious about aging and senility and hostile to the aged as other physicians and society in general. But throughout the twentieth century and especially by the mid-1930s, senile dementia was fast becoming a problem that psychiatrists simply could not ignore. The total number of people suffering from some form of dementia was relatively small in the first decades of the twentieth century.[1] But people with dementia were increasingly concentrated in the state mental hospitals, for which American psychiatrists were responsible. Faced with this reality, a growing number of psychiatrists—albeit still a small minority—began to study more rigorously the pathological processes involved in senile dementia.

In the first three decades of the twentieth century, senile dementia posed two sorts of problems. First, there were the difficult issues surrounding the classification and pathogenesis of senile dementia. What was its relationship to aging? To arteriosclerosis? To the newly created disease entity, Alzheimer's, first described in 1906? What was the relationship between the clinical symptoms of dementia and the various degenerative changes that had been associated with it? None of these questions would be resolved before World War II, and the issue of senile dementia's relationship to aging, as well as the pathogenesis of the disease, remains controversial to the present.

A second set of issues concerned the growing number of elderly patients with dementia who were becoming a disproportionately large part of the patient population of the state mental hospitals. Although psychiatrists in the second half of the nineteenth century abandoned the ideal of moral treatment as the cure for mental illness, the ideal upon which the psychiatric hospital as an institution had been founded, and by the 1930s began a nascent movement toward private psychiatric practice in the community, the profession as a whole remained rooted in institutional practice through World War II. Conditions within the mental hospitals continued to be an important issue for the authority and prestige of the psychiatric profession.[2] Because psychiatry regarded the aged insane as incurable and their behavior difficult to manage, their presence in large numbers in the mental hospitals was particularly troubling for the profession.

In the mid-1930s, a new theory of dementia emerged in American psychiatry as an answer to both sets of problems. Frustrated by the inability to resolve issues of classification and pathogenesis through the methods of clinical-pathological correlation and stimulated by the rising influence of psychoanalysis, a group of American psychiatrists forged a distinctly new approach to senile dementia that emphasized psychosocial factors over brain pathology in the etiology of dementia. This approach sidestepped the difficult issues regarding the classification and biological pathogenesis of the disease. Since psychosocial factors were thought to play an important role in all forms of dementia, precise categorization was of secondary importance. Similarly, the issue of the relationship between various forms of pathology in the brain and the pathogenesis and manifestation of the symptoms of dementia was no longer so pressing because psychosocial factors in the onset of dementia gave the clinician plenty to work with. Regarding the second set of issues, the psychodynamic approach to dementia provided a rationale for therapy and so cut through the therapeutic nihilism toward these patients that threatened to envelope the mental hospitals. Given the possibilities the psychodynamic approach opened up for a more satisfying understanding of dementia and for meaningful therapeutic intervention in a most difficult group of patients, it is no surprise that from the mid-1930s through the 1950s American psychiatrists were increasingly interested in age-associated dementia. In the ten years from 1926 to 1935, nine articles concerning senile dementia or Alzheimer's disease (or both) were published in the two leading professional journals of American psychiatry and neurology; in the following decade, thirty-six articles appeared.[3] Although most historical accounts of senile dementia and Alzheimer's disease have ignored this period because it did not produce any meaningful contributions to the understanding of brain pathology and genetics which so intensely occupies researchers today, many observers at the time thought the period one of great progress and optimism.[4]

"MORE PEOPLE OUTLIVE THEIR BRAINS": PSYCHIATRY, THE AGED PATIENT, AND THE CRISIS OF THE STATE MENTAL HOSPITALS

The surge in interest in senile dementia can be understood only in the context of American psychiatry's ongoing hostility toward aged patients, who were becoming a disproportionately large part of the patient population of the state mental hospitals throughout the first half of the century. Because psychiatry regarded the aged insane as incurable, their presence in large numbers undermined the therapeutic environment that the state hospitals were supposed to provide. Because the overall population was aging, the problem was regarded by many as an impending crisis—a demographic avalanche that would bury the state hospitals and the professional authority of psychiatry along with them.

The problem of aged patients in the state mental hospitals began with changes in state public policy around the turn of the century. In separate studies, historians Carole Haber and Gerald Grob have shown that through most of the nineteenth century psychiatrists were able to keep aged patients out of the mental hospitals by arguing that senility did not really constitute insanity—and certainly not curable insanity—and that senile old people thus could be cared for best in the homes of their families or in almshouses or private old-age homes (the nursing home industry as we know it today did not yet exist). Because almshouse care was the least expensive way of caring for aged patients who had to be institutionalized, local welfare officials also sought to keep the aged out of mental hospitals.[5]

At the end of the nineteenth century this changed. Both the absolute and proportional number of aged patients admitted to the state hospitals increased dramatically—a trend that continued through the early 1960s.[6] This did not mean that most or even many psychiatrists had concluded that senile dementia should be regarded as a form of insanity suitable for treatment in mental hospitals. Rather, as Grob persuasively argues, two public policy developments, the decline of the almshouse as a viable public institution and, more important, the decision by states to assume responsibility for all mentally ill persons, account for the influx of aged patients into the mental hospitals. As individual states enacted legislation that shifted financial responsibility for the insane to the state governments, local welfare officials reclassified their aged almshouse inmates as insane. Families often preferred to have their aged relatives admitted to the mental hospital rather than the almshouse if they could not be cared for at home. The care received in mental hospitals was widely perceived as superior to that in the almshouse; moreover, confinement of a family member to the almshouse remained a badge of shame and failure. The result of all this was that responsibility for the care of elderly people with dementia was increasingly thrust on psychiatry. Grob cites the attitude of Charles Wagner, superintendent of the Binghampton State Hospital in New York, as typical. In the New York State Commission of Lunacy's annual report for 1900, Wagner remarked:

> We are receiving every year a large number of old people . . . who are simply suffering from the mental decay incident to extreme old age. A little mental confusion, forgetfulness and garrulity are sometimes the only symptoms exhibited, but the patient is duly certified to us as insane and has no one at home capable or possessed of means to care for him. We are unable to refuse these patients without creating ill-feeling in the community where they reside, nor are we able to assert that they are not insane within the meaning of the statute, for many of them, judged by the ordinary standards of sanity, cannot be regarded as entirely sane.[7]

As Wagner's remarks indicate, whatever psychiatrists may have felt about the issue, they were not able to prevent the placement of aged patients in mental hospitals. Although comprehensive and reliable national data are impossible to reconstruct, Grob cited data from Massachusetts and New York that illustrate the magnitude of change. In 1855, the rate of first admission for male and female patients over age 60 to Massachusetts mental hospitals was 70.4 and 65.5 per 100,000; in 1939–41 the corresponding rates were 279.5 and 223 per 100,000. In 1919–21, 9.2 percent of first admissions to state hospitals in New York were aged 70 and over; in 1949–51, 26.2 percent were 70 and over. The percentage of the population in New York over age 65 was 4.7 in 1920 and 8.6 percent in 1950.[8]

From the 1930s through the 1950s, dire pronouncements about the burden that aged patients placed on psychiatry and the impending crisis of the state hospitals were an increasingly prominent feature of the professional literature of American psychiatry. "Our institutions promise to become in time vast infirmaries with relatively small departments for younger patients with curable disorders," Richard Hutchings cautioned in his 1939 address as president of the American Psychiatric Association.[9] Following up on Hutchings' remarks, the authors of a 1942 study of the senile and arteriosclerotic psychoses noted that the number of people 65 years of age and over in the U.S. population increased by

35 percent from 1930 to 1940, with a corresponding increase in admissions to mental hospitals. (Actually, the increase in admissions far outstripped the increasing proportion of aged persons in the population.) "So pressing is the problem," the authors argued, "that the character of mental hospitals is in danger of again reverting to a functional level of custodial care."[10] The authors of a 1942 study of senile and arteriosclerotic psychoses at the Pennsylvania Hospital began by remarking that "the increasing hospitalization of persons suffering from senile and arteriosclerotic mental disease lays on the medical profession a pressing obligation to study preventative and remedial measures by every means within its power."[11] The prominence of the problem in psychiatry during this period was summed up by former American Psychiatric Association president Abraham Myerson in 1944, when he argued that progress in longevity brought with it the "distressing fact that more people . . . outlive their brains." Psychiatry would "have to face this fact, since the real increase in mental diseases comes by the roads of senile dementia with its plaques, and cerebral arteriosclerotic dementia, with its more direct cardio-vascular changes."[12]

Psychiatry reacted to this in two ways. One, as the bulk of this chapter shows, was to redefine the problem in a manner that created new ways for psychiatrists to deal with the problem. But another way was to redouble efforts to create alternative policy solutions to the problem of caring for aged patients.[13]

Conditions in the mental hospitals became a public scandal in the 1940s and 1950s with the publication of a series of exposés by Albert Deutsch and other journalists.[14] Graphic, if not sensationalistic, images of what things were like could be found in the psychiatric literature as well. For example, a 1957 article describing the treatment of senile dementia with electroconvulsive therapy contained this description: "Some [patients with dementia] were completely withdrawn, refused to eat, tried to commit suicide or in a most morbid way played with or ate their excrements. Others were destructive to themselves or their surroundings, tearing up their mattresses, banging the doors, screaming at the top of their voices or attacking patients and attendants if not heavily sedated or in restraints."[15]

By the late 1950s, the need to find alternatives to the hospital for care of the demented elderly was a common theme in the psychiatric literature.[16] An alternative was created in the 1960s, when, through provisions in Medicare and Medicaid, the federal government assumed responsibility for funding nursing home care for elderly patients. The result was the transfer of many thousands of elderly patients with dementia out of the mental hospitals and into nursing homes and various community care arrangements, a change as rapid and dramatic as had been the earlier shift into the mental hospitals.[17]

NOSOLOGICAL TANGLES: THE CREATION OF ALZHEIMER'S DISEASE AND THE LIMITS OF PATHOLOGY

If the troubling proliferation of aged people with dementia in the wards of the state mental hospitals was a major impetus for increasing the attention psychiatrists paid to senile dementia during this period, the failure of an organic approach to solve a number of vexing issues regarding the classification and pathogenesis of age-associated dementias helped determine the particular shape that attention took—a psychosocial model of dementia. The principal architect of that model, David Rothschild, trained in both neuropathology and psychoanalysis, began studying senile dementia with the aim of establishing clinical-pathological correlations as Alzheimer and Kraepelin had, but he turned toward a psychosocial explanation of dementia when firm correlations could not be established.[18]

As we have seen, the drive of American psychiatrists to create a concept of senile dementia rooted in clinical-pathological correlation began in the late nineteenth century and was part of a broad transformation of psychiatry. In large part, this turn was fueled by a desire among psychiatrists to catch up with other branches of clinical medicine, which had surpassed psychiatry in prestige thanks in large part to their ability to define the pathogenesis and etiology of discrete disease entities through bacteriological and pathological research, even if this seldom led to effective therapies.[19] The discovery by German psychiatrists in 1857 that general paresis, one of the most common forms of insanity, was connected to syphilitic infection raised new hopes that clinical-pathological correlations would lead to etiological theories, and ultimately therapeutic interventions, for other forms of mental illness.[20]

Although these grand hopes never materialized, in the first decade of the twentieth century, German researchers led by Emil Kraepelin and his protégés Alois Alzheimer and Franz Nissl nearly made senile dementia the second major mental disorder for which a clear pathological basis had been established. But for all their undeniable brilliance as clinicians and neuropathologists, Kraepelin and his group were unable to firmly establish the boundaries of the disease because they could not resolve the fundamental issue of whether the clinical symptoms and pathological structures they described constituted a disease entity or a part of the normal processes of aging.[21] Ironically, Kraepelin—whose fame rests with replacing the nineteenth-century Babel of conflicting idiosyncratic classificatory schemes based on symptoms with a relatively simple and rational nosological system based on careful observation of the natural history of mental diseases— added to the confusion by creating the entity he called Alzheimer's disease to distinguish the relatively rare cases in which dementia developed before the age of 65 (pre-senile dementia) from the common occurrence of dementia in more advanced ages (senile dementia). Kraepelin made this distinction, in the 1910 edition of his tremendously influential textbook, *Psychiatrie: Ein Lehrbuch für Studierende und Ärtze,* on the basis of only a handful of reported cases of pre-senile dementia and even though the pathological hallmarks, clinical symptoms, and progression of both pre-senile and senile dementia were virtually identical. Age of onset appeared to be the only criterion on which the distinction was made.[22] Subsequent researchers had to puzzle not only about the relationship of senile dementia to the normal processes of aging but also about the relationship between Alzheimer's pre-senile dementia and senile dementia and whether the former was related to aging in the same way or even at all.[23]

Alzheimer first described the disease that Kraepelin would eventually name for him at a meeting of the South West German Psychiatrists in Tübingen in 1906. He presented a brief report of the case of a 51-year-old woman who developed progressive dementia, accompanied by focal signs (symptoms such as aphasia or apraxia which suggest damage to specific, localized areas of the brain), hallucinations, and delusions. On postmortem, her brain was found to be atrophied, and it contained numerous senile plaques and a newly observed pathological structure—densely twisted bundles of neurofibrils, or neurofibrillary tangles, which were made visible to microscopic observation through a newly developed silver-staining technique.[24] Most of the brief report is devoted to a description of her clinical symptoms and the pathological changes found in her brain at postmortem. But Alzheimer concluded the report by suggesting that "on the whole, it is evident that we are dealing with a peculiar, little-known disease process. . . . We must not be satisfied to force it into the existing group of well-known disease patterns."[25]

While this seems to suggest that Alzheimer did indeed believe that this case belonged in a new category, one distinct from senile dementia, his opinion was not entirely clear from the brief report. In any case, it is hard to see what could have been regarded as new about the clinical or pathological description of the case. The clinical symptoms Alzheimer described were essentially the same as those found in senile dementia which had been described often in the literature, though Alzheimer suggested, as others did in the years that followed, that some

of them—especially the focal signs—were more pronounced in pre-senile dementia. Senile plaques had been identified and associated with senile dementia in the
previous decade, and the new finding of neurofibrillary tangles was quickly demonstrated to be a feature of senile dementia as well.[26] The only significantly new
feature of this case was the young age of the patient. Was this enough to declare a
separate disease category, as Kraepelin claimed? In 1911, a year after Kraepelin's
textbook declared that it did, Alzheimer himself seemed to argue that it did not. In
a lengthy paper that included a more thorough report of the first case as well as a
second case and a discussion of a handful of similar cases of pre-senile dementia
reported by others, he wrote that "as similar cases of disease obviously occur in
the late old age, it is therefore not exclusively a presenile disease, and there are
cases of senile dementia which do not differ from these presenile cases with
respect to the severity of the disease process. There is then no tenable reason to
consider these cases as caused by a specific disease process. *They are senile psychoses, atypical forms of senile dementia.* Nevertheless, they do assume a certain
separate position, so that one has to know of their existence . . . in order to
avoid misdiagnosis."[27] As German Berrios concludes, it seems that all Alzheimer
meant to emphasize in describing these cases in both the 1907 and 1911 publications was that senile dementia could occur in a younger person.[28]

So why, then, did Kraepelin assert the existence of the new entity, Alzheimer's
disease? A number of speculative theories have been proposed: that he did so as
part of a struggle against the inroads being made by Freud and psychoanalysis,
hoping to create in Alzheimer's disease a second example (general paresis being
the first) of a mental illness with a clearly defined pathological substrate;[29] that he
did so in order to garner prestige for his department, which was in a rivalry with
that of Arnold Pick in Prague;[30] that he did so in order to justify the creation of
Alzheimer's expensive pathology lab in Munich;[31] finally, that he did so with full
intellectual honesty out of the assumption that differences in age of onset were a
sufficient reason to make the distinction and that, in any case, he was reserving
final judgment for more decisive evidence.[32] These explanations are not mutually
exclusive, of course, but as Berrios correctly points out, they are all highly speculative. Until more systematic historical work is done on Kraepelin and his professional world, his decision remains inscrutable.[33]

Whatever Kraepelin's reasons for creating Alzheimer's disease as an entity
distinct from senile dementia, it was generally accepted by psychiatrists in the
United States, though not without expressions of confusion and consternation.[34]
As described by Martha Holstein, American psychiatrists in the second and third
decades of the twentieth century were predisposed to accept Alzheimer's as a

distinct entity for several reasons. First, since the late nineteenth century, they had been enduring withering attacks from neurologists for a lack of scientific rigor in general and the inability to identify the pathological mechanisms of mental illness in particular. Alzheimer's disease seemed an answer to this charge. Second, although she does not develop the implications of this, she does point out that the two most prominent researchers involved in writing about Alzheimer's disease— Alfred Barrett and Solomon Fuller—had both spent time studying with Alzheimer in Munich and were thus inclined to approach the problem in the same way.[35] But the most important factor, Holstein argues, was that, like Kraepelin, American psychiatrists simply found the logic of age classification too compelling to ignore. Prevailing negative views of old age made it hard to distinguish between the normal and the pathological in senile dementia. Dementia in old age was regarded as a common, if not expected, occurrence in old age, whereas dementia occurring at earlier ages seemed more clearly the result of a distinct pathological process. It thus seemed logical and useful to regard the latter as a disease, while the former remained in an ill-defined position between pathology and normal aging.[36]

Regardless of their motivations, support for the distinction among American psychiatrists was tepid at best. Solomon Fuller, whose role in the early history of Alzheimer's disease has been long overlooked, summed up the situation in two important articles published in 1912.[37] Fuller was the first African American psychiatrist in the United States, born into one of the wealthiest and most prominent families in Liberia. He came to the United States in 1889 to attend college and medical school at Boston University. During an internship at the Westborough State Hospital for the Insane near Boston he became interested in neuropathology and continued this interest with graduate study in Germany, where he took a position as an assistant in Alzheimer's lab in 1904.[38] In 1905, Fuller returned to Westborough, where he continued his research on Alzheimer's disease and other conditions. In his first published article on Alzheimer's disease, Fuller noted that work in this field was still very new, and he hedged a bit as to whether Alzheimer's was in fact a disease entity: "While more or less definite mental symptoms and structural alterations are referred to in this paper, the recorded cases are too few—even those showing important variations—to warrant maintaining anything comparable to the paradigm of general paresis"; he followed with a list of some of the variations in the clinical and pathological symptoms that Alzheimer brought together in his 1907 description—for example, a case with numerous large plaques but no tangles, a case with tangles but no plaques, and variations in the manifestation of focal signs and other clinical symptoms. Nonetheless, despite these variations and the small number of cases on which to stake the claim, Fuller thought that, overall, "clinical and anatomical findings offer a striking similarity" that made "the assumption of a clinical type or subgroup . . . not altogether unwarranted." He framed his case report of a 56-year-old man with dementia as "cumulative data toward the isolation of a type which while lacking at present some of the postulates of a disease entity, may yet crystallize into such."[39] If these comments constitute acceptance of the new category, it could not be more tentative. In the second article, he and his colleague Henry Klopp explicitly accept Alzheimer's disease as an entity, but more as a formal nod to Alzheimer's contribution than as a meaningful distinction between Alzheimer's disease and the larger field of senile dementia: "If, then, these are cases of atypical senile dementia, the question could fittingly arise why a special designation—Alzheimer's disease—since after all, they are but part of a general disorder. Still, the profession must remain indebted to Alzheimer for having first called attention to this type of cases [sic]." Fuller and Klopp then noted that Alzheimer himself in his 1911 paper did not claim Alzheimer's disease to be a distinct entity but rather "representative of a senile psychosis—atypical senile dementia," a view the authors shared not only with Alzheimer but with several other European contributors as well. The authors commented no further on the issue in this article but seemed to acknowledge that the term had come into accepted usage. "The case which forms the subject of this paper is, in our opinion, an example of the group now designated as Alzheimer's disease."[40]

Fuller's ambivalent acceptance of Alzheimer's disease as a distinct disease entity was echoed by other researchers in the decades that followed. For example, in a 1929 article, William Malamud and Konstantin Lowenburg argued that their own findings and the preponderance of evidence in the literature suggested that Alzheimer's disease was not limited to the pre-senium, but they nonetheless thought that the distinction should be maintained on pragmatic grounds. Since little was actually known about senility, saying that Alzheimer's disease was a form of senile dementia would not add anything to the understanding of Alzheimer's but would blur the meaning of senility by linking it to conditions occurring in earlier ages.[41] On the other hand, in 1936, David Rothschild and Jacob Kasanin argued that practical considerations dictated the opposite: since the pathological pictures were so similar, advances in understanding Alzheimer's would have the practical benefit of shedding light on the larger problem of senile dementia. "It is evident," they concluded, "that in a broad discussion of Alzheimer's disease one must include also the problems of senility in its normal and pathologic aspects."[42] The distinction between Alzheimer's disease and senile dementia maintained its ambiguous position in official nosology until the mid-1970s, but from the outset it was a distinction with little difference. Officially, Alzheimer's disease remained a rare form of pre-senile dementia, and that is why the term slipped into obscurity over the years. But every researcher of this period seriously engaged in the study of dementia was aware that Alzheimer's disease and senile dementia were for all practical purposes the same entity. Since very little was at stake in a field that remained so undeveloped, they simply found no compelling reason to abandon the distinction that Kraepelin had made.[43]

In his 1911 paper, Alzheimer followed up his disavowal of Alzheimer's disease with a call for "future research to collect a larger number of such cases, which, as the observations of this department show, should not be too rare in order to establish the symptomatology of this group even more clearly, and to substantiate their position with respect to senile dementia on an even firmer basis by proving the existence of further transitional cases."[44] Fuller, too, suggested that there was "lively interest shown in the type of mental disorder to which Alzheimer was the first to call attention" and that "these cases clearly indicate psychoses occurring in or about the period of the senium are a rich field for clinical and anatomical research."[45] But if there was "lively interest" in these cases early in this decade, it soon dissipated; although this untilled field appeared rich and fertile to Alzheimer, Fuller, and other pioneering neuropathologists of this era, it could fairly be described as barren in the early 1930s. A few advances were made in understanding the pathology of age-associated dementia during the 1920s which would become important to researchers in the 1970s, such as Paul Divry's identification of the senile plaque as composed of amyloid protein, Friedrich Struwe's observation of senile plaques in the brains of young patients dying with Down's syndrome, and the observation by several researchers of familial clusterings of Alzheimer's disease.[46] But two decades of sporadic research had failed to resolve any of the fundamental issues about the boundaries of senile dementia.

In addition to the problem of the relationship between Alzheimer's disease and senile dementia, there was the problem of arteriosclerosis. Although many researchers made a careful distinction between senile (Alzheimer-type) dementia and cerebral-arteriosclerotic psychoses, the traditional view that arteriosclerotic processes were somehow causative of all dementia in later life continued to appear in the literature, and the diagnosis of arteriosclerosis seems to have been greatly overused through the 1960s.[47] More broadly, the problem of whether to consider senile dementia a physiological or pathological process, a product of aging or of disease, remained unresolved. As we shall see in chapter 4, the issue remains vexing to the present day.[48] By the early 1930s, research into senile dementia and Alzheimer's disease had established an additional anomaly: in several large autopsy series, the correlation between the clinical symptoms of dementia and the presence of brain pathology postmortem was surprisingly loose. In some cases, the senile plaques and neurofibrillary tangles that were found in the brains of people with dementia were also found in the brains of people who had shown no sign of dementia in life. In other cases, the brains of people who died while they were in advanced stages of dementia were found at autopsy to be relatively intact. With these issues unresolved, the boundary between aging and disease (or diseases) remained unclear.

"FACTORS OF A MORE PERSONAL NATURE": DAVID ROTHSCHILD AND THE PSYCHODYNAMIC MODEL OF SENILE DEMENTIA

In grappling with these issues, Massachusetts psychiatrist David Rothschild was moved ultimately to look beyond the deterioration of brain tissue to explain dementia. In so doing, he developed an etiological theory that was so compelling that issues of boundary and classification which had plagued the field for years receded into the background. If one could look to a set of factors that explained dementia whether it appeared in the pre-senile or senile period, whether it was accompanied by Alzheimer-type pathology, arteriosclerotic lesions, both, or neither, the inability to resolve these issues seemed far less troubling.

Few biographical details about Rothschild are available. He spent his entire career in the Massachusetts state hospital system, first as research director of the Foxborough Hospital from 1927 to 1941 and then as clinical director at the Worcester Hospital from 1946 to 1956, publishing nearly thirty papers on the psychiatric problems of aging. Rothschild began his career by trying to resolve some of the anomalous findings concerning the pathology of Alzheimer's disease and senile dementia which had been accumulating in the decades after Alzheimer's initial description. In papers published in 1931 and 1936, Rothschild and his colleagues argued that, since the pathological structures characteristic of Alzheimer's disease and senile dementia had been found in a variety of other conditions such as spastic paralysis, hyperthyroidism, and multiple sclerosis, these structures were not an expression of any one disease process but an indication of a special type of tissue reaction that might occur in response to a variety of biological factors.[49] Some researchers pursued this idea, while others sought a single cause for the tissue changes.[50] Rothschild and his colleagues, however, began to explore the relationship between biology and personality.

In their 1936 paper, Rothschild and Jacob Kasanin explained the severity of emotional and personality disturbances in Alzheimer's disease—which they claimed were typically more pronounced than those found in senile dementia—as the attempts of a relatively strong organism to compensate for the oncoming disease. The more severe clinical symptoms of Alzheimer's disease could be understood as "an attempt at rejuvenescence in the face of an abnormal aging process."[51] When this idea was considered in light of research demonstrating that the brains of some aged people who had died with normal mentality contained the same pathological structures as those of people with dementia, the possibility of a strikingly new conception of senile dementia was apparent.

In a flurry of articles published in the major psychiatric journals between 1937 and 1952, Rothschild reported that his own clinical and histological investigations verified a discrepancy, first noted by the German researcher N. Gellerstedt, between the presence and degree of dementia in the living patient and the presence and degree of pathological structures found in autopsy.[52] In accounting for this discrepancy, Rothschild rejected an available biological explanation: a German researcher, A. von Braunmühl, argued that the pathological structures of senile dementia were actually secondary phenomena of other, undetectable senile changes. It was thus possible to say that "the brain of a senile patient is more severely damaged than that of a patient of normal mentality, even though the demonstrable alterations may be the same in both cases." The problem with Braunmühl's explanation was that it could be reversed with equal justification: if the real senile changes could not be detected, then it could be argued that the brain of a person of normal mentality was more profoundly damaged than the brain of a person with severe dementia. Braunmühl's theory, Rothschild argued, "is effective only if it is applied in one direction, a direction determined by preconceived ideas as to what tissue damage should be. Therefore, it cannot be accepted as a satisfactory explanation of the[se] troublesome discrepancies."[53]

Finding a satisfactory explanation moved Rothschild away from what he saw as biological reductionism. "Too exclusive preoccupation with the cerebral pathology," he argued, had "led to a tendency to forget that the changes are occurring in living, mentally functioning persons who may react to a given situation, including an organic one, in various ways."[54] The lack of correlation between clinical and pathological data could best be accounted for by a differing ability among individuals to compensate for organic lesions. An individual's ability to withstand organic damage was decreased both by personality defects and by stress and life crises.[55]

Rothschild was not arguing that pathological structures were unimportant.

There were "limits beyond which senile lesions will, no doubt, produce a mental disorder in any person." But within a fairly wide range, he argued, they did not lead inevitably to mental illness. Structural alterations were always present, but their significance could be understood only in light of "factors of a more personal nature." Each case, he concluded, required "individual scrutiny, and instead of focusing attention solely on the impersonal tissue process or on more personal influences, the main object should be to estimate the relative importance of these two sets of forces as factors in the origin of the psychosis."[56]

THERAPEUTIC INITIATIVE IN THE STATE HOSPITALS

There was nothing inherently new in Rothschild's work, of course. It clearly reflected the dominance of psychodynamic theory in American psychiatry from the 1920s through the early 1960s, and particularly the psychobiological approach of Adolph Meyer, which conceived of all mental phenomena as a dialectical interplay between biological, social, and psychological forces.[57] As Rothschild pointed out, his explanation of senile dementia was, "in fact, the attitude of modern American psychiatry to mental disorders in general."[58] Applied to senile dementia, however, this attitude had the effect of challenging the therapeutic nihilism with which psychiatrists habitually viewed older patients. As Elvin O. Semrad put it in the discussion that followed one of Rothschild's papers published in 1947, "the emphasis on personality factors helped lead me and my associates out of the convenient balm for our frustrations as therapists which preoccupation with the neuropathologic damage and internal medical aspects of these problems allowed."[59]

In the 1940s and 1950s, there was a surge of interest in therapies, both biological and psychodynamic, which had previously been considered inappropriate for aged patients—reflecting both the optimistic avenues that Rothschild's work opened and the desperate empiricism to which psychiatrists were driven by the crisis of the mental hospitals. From 1935 to 1959, thirty-five articles appeared in the two leading professional psychiatric journals discussing therapy for patients with dementia; in the fifteen preceding years, there had only been *three*. Many of these articles cited Rothschild or related work, and even those that didn't implicitly recognized the changed terrain on which their therapeutic efforts were based. A 1955 article on the use of electroconvulsive therapy on aged patients was typical: following the usual litany concerning the growing importance of geriatric psychiatry given the "mounting numbers of patients admitted to the senile wards of mental institutions," the authors noted that "considerable change has taken place in our thinking about the subject" over the previous decade, and they then went on to cite Rothschild and other proponents of the psychodynamic model.[60] Even treatments that seemed to be working on the assumption of a biological mechanism—for example, vasodilators aimed at increasing cerebral blood flow—had an implicit psychodynamic aim. Improved brain function that psychiatrists hoped would result from increased cerebral blood flow would make the patients more alert and able to undergo psychodynamic treatment.

Increased interest in therapeutic interventions in the care of elderly people with dementia was accompanied by a heightened interest in developing clinical procedures for evaluating the extent of organic damage in senile dementia. Many psychiatrists stressed the need to differentiate carefully between irreversible organic conditions and dementias that were produced by reversible conditions such as nutritional insufficiency or drug toxicity. Hospitalization itself, psychiatrists recognized, could be a source of dementia because of the disorienting effects of confinement in a strange location, disruption of routine, and the danger of over-medication.[61] Implicit in the literature was the belief that senile dementia was drastically overdiagnosed. J. Luhan argued in 1945 that it was common practice in the state hospitals to "accept without staff presentation or review, the diagnosis of psychosis with cerebral arteriosclerosis or senility made on the patient's admission." In the same article, Meyer Solomon argued that this practice often destroyed the morale of the patient and family without foundation. The term "senile dementia" ought to be dropped, he thought, in favor of "geriopsychosis," which would reflect the variability of origins and prognosis in psychotic conditions of the elderly. In any case, psychiatrists would need "to make a careful re-evaluation of our attitudes toward old people and of the too hurried diagnosis and management of their mental disorders."[62]

It is difficult to evaluate the effectiveness of the therapeutic initiatives undertaken during this period. The growing literature on treatment, while generally optimistic in tone—and remarkably so given the long-standing psychiatric pessimism regarding senility—reported varying degrees of success depending on the therapies and goals in question. Electroconvulsive therapy (ECT), aimed at lessening the confusion of patients with dementia and restoring their ability to function socially, produced the most favorable results. The previously quoted article followed 112 patients treated with ECT; it reported that 78.5 percent were able to be discharged from the hospital and that 55 percent of these were able to stay out of the hospital for more than a year.[63] Other studies reported similar levels of success.[64] Drug therapies and group and interactive therapies aimed at calming agitation and making elderly patients with dementia more manageable on the wards also received generally favorable reports.[65] Vitamin and other nutritional therapies and hormonal and drug treatments aimed at preserving or restoring the intellectual capacities of elderly, cognitively impaired patients had much more ambiguous results.[66] But all these studies were highly problematic when judged by contemporary standards; few of them were blinded or randomized, and all were plagued by problems and ambiguities and inconsistencies in diagnosis and the conceptualization of dementia. In any case, none of the positive findings for these various treatments were replicated by more careful studies conducted in the 1970s, and none are currently regarded as efficacious in the treatment of senile dementia.

Nonetheless, as Jack Pressman and Joel Braslow have separately pointed out in their extensive studies of lobotomy, wet packs, ECT, and other biological therapies employed in the state mental hospitals before the advent of chlorpromazine, there is a distinction to be made between efficacy and effectiveness.[67] Efficacy is measured by how well a treatment performs in clinical studies, while effectiveness is determined by how well a treatment does in everyday practice. With the exception of ECT, none of these treatments used in the state mental hospitals for a variety of mental illnesses in the first half of the twentieth century meet contemporary standards for efficacy. But they were clearly effective treatments within the context of workaday psychiatric practice in the state mental hospitals. As Braslow puts it, "whatever biological consequences a particular therapy has are mediated by and interpreted through the way doctors see disease and its cure, which are themselves determined by therapeutic practice. . . . The 'biological effects' of a treatment, though obviously important, assume a less important role in understanding the effectiveness of a particular remedy" than the context in which they were used.[68]

Looked at this way, the therapeutic initiatives that appeared in the 1940s and 1950s can be understood as at least modestly effective. For institutional psychiatrists struggling under the burden of caring for these patients, any therapeutic success, no matter how limited, was clearly welcome. But regarding the effect of these therapeutic efforts on conditions within the state mental hospitals, one must balance the frequent optimistic pronouncements that the state hospitals were improving with equally frequent pronouncements about their deterioration and the impending crisis. Conditions in the mental hospitals could vary significantly from state to state. Given the realities of overcrowding, understaffing, and underfunding in many state hospitals, the therapeutic initiatives described in this section were likely to have affected only a relatively small number of patients at the few state hospitals that were in a position to try them. Conditions at most state hospitals could hardly be expected to improve, whatever sorts of patients were crammed into them, unless more basic problems of staffing and funding were resolved.

OBJECTIONS TO THE PSYCHODYNAMIC MODEL IN AMERICAN AND BRITISH PSYCHIATRY

Although Rothschild's psychodynamic model was clearly dominant among American psychiatrists writing on senile dementia through the 1950s, a strict organic approach had never completely disappeared.[69] A handful of articles appeared in the leading psychiatric journals in the 1940s and 1950s which explained dementia on the basis of biological changes alone, but only one—an ambitious study by Meta Neumann and Robert Cohn—addressed Rothschild's theory directly. Neumann and Cohn studied 210 cases of dementia in a state hospital and asserted firmly that Alzheimer's disease and senile dementia were identical and that the distinction ought to be dropped. They acknowledged the discrepancies cited by Rothschild between the degree of dementia found in the living patient and the degree of pathological damage found at autopsy, but because the overwhelming majority of patients with dementia showed large numbers of plaques and tangles at autopsy, they concluded that the link should be considered definite. A scattering of anomalous cases was not sufficient grounds for dismissing the evident organic basis of the disease.[70]

Objections to Rothschild's work were much stronger in Britain, where the psychodynamic model of dementia never caught on. This was not surprising, of course, because British psychiatry in general was far less enamored of psychoanalysis and psychodynamic psychiatry. I have found only one example of British work of this period in full agreement with Rothschild's view. Felix Post concluded that his own study "in conjunction with the findings of other workers [Rothschild and other American psychiatrists] indicates that factors other than physiological and anatomical ones are responsible for the origin of senile and arteriosclerotic mental disorders."[71] More typically, the British psychiatric literature of the 1940s and 1950s adopted a weak version of Rothschild's argument when discussing the discrepancy between pathology and dementia: it maintained that differences in the physiological capacities or "constitution" of individuals accounted for differences in their ability to withstand a given amount of cerebral damage without developing dementia, but it simply ignored Rothschild's claims about psychosocial factors. Thus William McMenemey, citing the 1937 paper in which Rothschild first articulated the psychodynamic model, referred to Rothschild's "important suggestion that the ability of individuals to compensate for brain damage varies considerably and that this difference in reserve power of the brain will account for the fact that some elderly persons afflicted with organic disease of the brain develop senile dementia and others do not." McMenemey's phrase "reserve power of the brain" serves to limit the differing compensatory capacities of individuals to the physiological realm. Social, emotional, and psychological factors— which were at the heart of Rothschild's concept—were in McMenemey's view secondary phenomena that affected the manifestation of the real disease. "Once can imagine," McMenemey continued, "that the senile decorticating process has already begun and the patient awaits only an emotional shock or an attack of influenza to precipitate the onset of symptoms."[72]

Other critics were less coy. In a 1948 article much cited by contemporary researchers for forcefully arguing that Alzheimer's disease and senile dementia were the same entity, R. D. Newton argued that there was not an observable correlation between dementia and the compensatory and noncompensatory personality types Rothschild described, so he found little reason to consider psychological factors in the etiology of dementia.[73] German émigré William Mayer-Gross lauded Rothschild's work demonstrating the discrepancy between brain lesions and psychotic symptoms, but he distinguished between Rothschild's organic explanations, which he approved of, and his psychodynamic explanations, which he scorned. In a paper on arteriosclerotic dementia, for example, Rothschild had argued that greater cardiovascular resistance in some patients, as measured by cardiac weight, might in part account for these discrepancies—a hypothesis that Mayer-Gross found persuasive. But Mayer-Gross found it "difficult to follow him further when he eventually linked up cardiovascular with psychological resistance of the personality, thus paying his tribute to the 'psychobiological' trend in present-day American psychiatry."[74]

The most important reaction to Rothschild in Britain, however, came from Martin Roth, who in the early 1950s began what would be a sustained and rigorous attack on the psychodynamic model, particularly on the claim that the plaques and tangles were unrelated to the clinical manifestation of dementia. Roth developed his interest in geriatric psychiatry in training with Mayer-Gross at the Maudsley and later the Crichton Royal Hospital in Scotland.[75] In the 1950s, Roth was interested in establishing a more rigorous classification of the mental disorders in later life and carried out a large-scale study of the clinical histories of 450 older patients admitted to Graylingwell mental hospital, where he had taken the position of clinical director. On the basis of a thorough review of patient records and follow-up examinations, Roth divided the patients into five categories of mental disorder: affective, late paraphrenia, senile, arteriosclerotic, and acute confusional. The first two disorders Roth regarded as functional and the last three organic. Roth found significant differences in the outcomes in terms of mortality and morbidity between all the categories, but in particular between the functional and organic groups. The affective and paraphrenic groups lived significantly longer and had a much better prognosis than the senile and arteriosclerotic groups. The acute confusional group occupied an ambiguous middle position in between, with a relatively high level of recovery compared with the senile and arteriosclerotic groups but a high degree of mortality compared with the affective and paraphrenic groups. These sharply differing outcomes, Roth argued, indicated that these disorders had unique natural histories and so ought to be regarded as distinct disease entities to be carefully diagnosed and treated.[76] Significantly, Roth found that the functional disorders accounted for almost 50 percent of the admissions to Graylingwell, a rate that was in line with reports from other hospitals in the United Kingdom. But in the United States, Roth noted, senile dementia and cerebral arteriosclerosis were reported at far higher levels, accounting, it seemed, for virtually all mental disorder in old age. "One cannot help wondering," Roth asked, "whether underlying the American figures there may not be a much wider interpretation of senile and arteriosclerotic psychosis than can be justified on theoretical and practical grounds." Careful differential diagnosis was crucial, Roth argued, because a "failure to make a correct diagnosis may be fraught with dire consequences for an old patient. . . . It need hardly be said that a fatal outcome, after many months or years in hospital with some preterminal physical decline and mental confusion, is no confirmation for a diagnosis of organic psychosis." In cases in which differential diagnosis was particularly difficult, Roth advised administration of electroconvulsive therapy as a definitive diagnostic test. If improvement occurred, it was safe to conclude the patient had a functional disorder, since "there is no evidence to suggest that where a cerebral atrophic process and not an affective psychosis is the underlying cause it is accelerated by such treatment."[77] Clearly Roth did not take seriously the reports from American psychiatrists that ECT had been an effective treatment for cases of senile and arteriosclerotic psychosis.

Yet it was not until the late 1960s that Roth, teamed with psychologist Gary Blessed and pathologist Bernard Tomlinson, thoroughly dislodged Rothschild's influence on American psychiatry and helped to put brain pathology back on the center of the stage—a story that will be told in chapter 4. By that time, Rothschild's influence was waning for other reasons, as was psychodynamic psychiatry in general. But through the 1950s, Rothschild's work remained very influential both within psychiatry and beyond. In large part, this was because of his theory's expansiveness: not only did the theory successfully explain the puzzling discrepancies between pathology and clinical dementia, but it could, it was argued, provide insight into the entire experience of aging. Following World War II, Rothschild and the psychiatrists who followed him seldom even undertook research into brain pathology. Instead, they increasingly thought of modern social relations as the pathology of senility.

CHAPTER THREE

From Senility to Successful Aging

By the mid-1930s, and much more intensely following World War II, middle-class discourse on old age was reshaped by the emergence of gerontology and related endeavors as a field of research and practice for a diverse array of bio-medical and social scientists, policymakers, activists, and entrepreneurs interested in the "gray market."[1] These various groups, only loosely allied and at times in conflict with one another, shared an optimistic attitude toward aging and a commitment to improving the lives of the elderly—what I term the gerontological persuasion.[2] Although gerontologists in the various disciplines worked within different theoretical frameworks and did not always agree on concrete matters of policy, they shared the belief that the modern sciences would be able to extricate healthy or "successful" aging from the pathologies that so often and, in their view, so needlessly accompanied it. The gerontological persuasion began with a broad concept of senility as one of the fundamental problems of aging, but it ultimately drove gerontologists and geriatricians to more rigorously define and demarcate the boundaries of normal aging and disease. These efforts, in conjunction with broader efforts to improve the situation of the aged, laid the foundation for the emergence in the 1970s of Alzheimer's disease as one of the most frightening and devastating of diseases at both the personal and societal level.

The 1939 publication of Edmund V. Cowdry's multidisciplinary handbook *Problems of Ageing*, and the 1937 Woods Hole conference on which it had been based, marked several important new developments. First, the collection of expertise from various fields was unprecedented, including some of the best-known scientists in the biological, medical, and social sciences; second, participants in the conference and contributors to the book and its subsequent two editions began to identify themselves as a research community and worked to develop more extensive professional networks focused on the problems of aging; third, and perhaps most important, generous funding from the Josiah Macy Jr. Foundation marked the beginning of an infusion of significant philanthropic and governmental financial support for systematic research on aging.[3]

During the 1940s and 1950s, the organizational infrastructure of the aging field began to take shape as some of the most important professional associations and research institutes in this field were established. Major professional associations founded in this period include the American Geriatrics Society (1942), the Gerontological Society of America (1945), the National Council on Aging (1950), and the American Society on Aging (1954).[4] The number of government or university-based centers or laboratories conducting aging research also grew dramatically during this period: Nathan Shock's *Trends in Gerontology* listed seventeen such programs in its first edition (1949), but the number had grown to forty-five by the second edition (1957).[5] Shock's own Gerontology Research Center in Baltimore represented a significant increase in the federal government's commitment to aging research. Established as the Gerontology Unit of the National Heart Institute in 1941, it grew under thirty-five years of Shock's energetic leadership from virtually a one-man operation into a program that, by 1975, supported the work of 175 scientists when it became the Intramural Program of the newly established National Institute on Aging.[6]

As a result of these developments, both the amount and the breadth of published research on aging grew tremendously. An exhaustive bibliography of biomedical and social science research on aging between 1954 and 1974 produced more than 50,000 citations, and it was estimated that the amount of scholarly literature on aging published between 1950 and 1960 equaled what had been published in the previous 115 years.[7] Researchers entering the field in the 1940s and 1950s opened the investigation of virtually every aspect of human aging—from basic biological mechanisms at the cellular level to the broadest psychological, economic, political, social, and cultural aspects of the aging process.

Although the emergence of gerontology as a multidisciplinary field of study and the ripening of geriatrics as a well-defined medical specialty clearly marked the beginning of a new era in the social and cultural history of aging in the United States, these developments were just as clearly connected to questions and concerns about the place of old age in the modern world which had been widely debated in the United States since the 1870s. This connection could be seen particularly clearly in the discourse of geriatricians and social gerontologists, who sought to create a new, more optimist image of old age within the domain of medicine and American society as a whole. As with earlier discourse on old age, the gerontological persuasion was primarily concerned with the fate of the aged man in modern society, but it reframed the assumptions about the causes of senility. As described in chapter 1, from the late nineteenth century on, the dullness found in the senile, their isolation and withdrawal, their clinging to the past and lack of interest in worldly affairs were characteristically represented as the *symptoms* of senility—the social stigmata of the inevitable deterioration of the brain. Following World War II, gerontological discourse typically represented these as the *causes* of senility. The locus of senile mental deterioration was no longer the aging brain but a society that, through mandatory retirement, social isolation, and the disintegration of traditional family ties, stripped the elderly of the roles that had sustained meaning in their lives. When elderly people were deprived of these meaningful social roles, when they became increasingly isolated and were cut off from the interests and activities that had earlier occupied them, not surprisingly their mental functioning deteriorated. The elderly did not so much lose their minds as lose their place.[8]

Social gerontology framed senility within a narrative of ironic progress in which economic and social development brought new, and perhaps more daunting, problems. Senility was an excellent example of this because it was made possible only by the tremendous economic abundance and social progress that the nation enjoyed after World War II. The senile old man, so palpably lost in the modern world, was lost first because science had given him long life and second because economic productivity had made his labor superfluous and so robbed his life of meaning. Senility was no longer a problem of productivity—the failure of the aging man to measure up to the grinding demands of work in an industrial age—but a problem of leisure, his failure to create a meaningful life in the abyss of empty, unstructured time in which he found himself. The image of the senile man thus continued to be freighted with middle-class anxiety about the coherence and stability of the masculine self, but a self that now had to be established and maintained less through work than leisure.

Anxiety about this development was a major theme among post–World War II American intellectuals. As historian Wilfred M. McClay describes it, works as diverse as William H. Whyte's *The Organization Man*, David Riesman's *The Lonely Crowd*, Vance Packard's *The Status Seekers*, Herman Wouk's *The Caine Mutiny*, John Kenneth Galbraith's *The Affluent Society*, C. Wright Mills' *White Collar*, and Daniel Boorstin's *The Image* shared an indictment of modern American culture: "Americans, who had formerly prided themselves on their stubborn independence and devotion to the work ethic, had succumbed to a collectivism of the mind, a dangerous susceptibility to the mass appeals of advertising and media, a deification of organizational life, and a social ethic that relinquished personal integrity and authenticity and instead enshrined as the American Way the bland and anxious conformism of the new white-collar personality-selling occupations, as well as the leisure-and-consumption-oriented lifestyle of the sprawling new American suburbs."[9] If self-creation through work was blocked by the conformist imperative of corporate bureaucracy, self-creation through leisure, a leisure dominated by a mass consumer culture, seemed equally stultifying.

Discourse about senility was fueled by these anxieties. In the late nineteenth century, discourse on aging represented the senile man as the epitome of failure because his decaying brain collapsed under the pressure to produce in a modern industrial society. In the mid-twentieth century, the senile man continued to epitomize failure, but now it was not his aging brain that was the problem but his failure to adjust to a consumer-oriented leisure society. Deprived of meaningful work, the aging man needed to find in leisure activities the means of creating a self. Senility was the failure to accomplish this. Focusing attention on the problem of senility dissipated middle-class anxiety about the problem that every American faced in creating a meaningful, coherent, and stable life through practices of consumption and leisure. To the gerontological persuasion, and to much of American society, senility was thus an increasingly intolerable affront to the dream of progress, of the limitless acquisition and enjoyment of health, wealth, and personal fulfillment. The diverse academic, medical, social service, and business professionals who embraced the gerontological persuasion passionately attacked the social pathology of senility. This attack involved improving the material circumstances of old age through increasing public and private pensions, abolishing mandatory retirement, establishing a network of social and recreational services, and, perhaps most important, replacing the negative image of senility, which generated fear of and hostility toward the elderly, with positive images of successful aging, which generated the optimistic attitude necessary for the individual and society to meet the challenges of aging. Geriatrician Martin Gumpert, one of the most energetic and visible proponents of the gerontological persuasion, writing scores of articles on aging in popular middle-brow magazines and journals as well as several books, argued that the crucial factor was for people to realize that freedom from senility was possible.[10] "If people are asked whether they would like to live for a hundred or a hundred and twenty or a hundred and fifty years, most of them seem horrified. Why?" Gumpert wondered. "They are frightened by the grotesque ugliness, the obscenity, the misery of senility, the hell in which so many beloved ones languish before they die. They look with fear into the mirror that reveals the first symptoms of old age and they would give their souls to arrest its ravages." Gumpert emphatically believed that senility could and would be eradicated. "There are facts to prove it, methods to follow and hopes that can be realized." People deserved a better fate, "a more serene, more entertaining, more satisfying kind of life," Gumpert concluded. And they would have it. "All the technical means for creating happiness are available today. We need only change the direction of our energy from destruction to reconstruction."[11]

According to several historians, much of the post–World War II gerontological program had been accomplished by the end of the 1970s, creating what Peter Laslett has called the "emergence of the third age" and others have called the "young old."[12] The material circumstances of old age had markedly improved, significant legal protections had been won against age discrimination in the labor market, negative stereotypes were challenged, and the elderly themselves organized for political advocacy and action on their own behalf.[13] By recasting aging as normally a period of continuing health and activity, the gerontological persuasion made dementia stand out much more clearly as a pathological condition. Thus, if by the 1970s senility had not in fact been eradicated, it had at least been domesticated—relegated to various discrete disease entities that, at least in theory, no longer contaminated the entire experience of aging. The image of the successful elder—healthy and attractive, fully engaged with the world—had become at least as prominent in popular and professional discourse on aging as the stereotype of senility.

Whether Americans have paid for these accomplishments with their souls, as Gumpert had asserted they were willing, remains an open question.

"THIS SITUATION IS MAN'S FAULT": THE NARRATIVE OF IRONIC PROGRESS

The idea that senility was a product of civilization was, of course, not new to the gerontological persuasion. It was central to the modern stereotype of senility that emerged in the late nineteenth century. George Miller Beard, for example, had argued that the rapid pace of modern life, especially for middle-class "brain workers," produced nervous exhaustion and that senility was the ultimate wearing down of the mental machinery under the stresses of modern life. He saw no tragedy in this, concluding simply that progress rightly favored the young. The only problem was breaking society of the habit of thoughtlessly revering the aged for their wisdom when the intersection of biology and progress meant that the aged were anything but wise.

For several reasons, such a conclusion no longer seemed tenable to those working in the aging field in post–World War II America. In the first place, the aging of the population made such an attitude appear wasteful and inefficient, hardly the rightful outcome of progress. Consigning the fastest-growing segment of the population to "corner rocking chairs, to lifeless rooming houses, and even to mental hospitals" was increasingly seen as a major threat to the future prosperity of the nation.[14] In the second place, medical research into the processes of aging challenged the assertion of inevitable decline. A small but growing number of geriatricians were arguing that most of the chronic infirmities that beset old age were to some degree amenable to treatment and in many cases even preventable.[15] More important, as the previous chapter showed, psychiatrists had demonstrated that there was reason to doubt that mental deterioration in old age could be completely chalked up to the deterioration of the brain. To the gerontological persuasion, senility was an embarrassment to civilization and an impediment to progress. "This situation is man's fault," Gumpert argued. "Old age and senility, nowadays almost identical, are no more necessarily related than infancy and rickets. With our aggregate of medical knowledge no child need suffer from rickets, and no elderly person need suffer from senility."[16]

In 1945, Yale anthropologist Leo W. Simmons presented the clearest articulation of the narrative of ironic progress in his widely influential book *The Role of the Aged in Primitive Society*.[17] Synthesizing and statistically analyzing ethnographic literature on seventy-one preindustrial cultures from around the world, Simmons demonstrated that different types of cultures produced different attitudes toward and roles for the aged. Although Simmons' book was ostensibly concerned only with "primitive" cultures, its implicit comparison of the exotic opportunities for fulfillment open to the elderly in these cultures with the dead end of modern old age constituted a powerful critique of the treatment of the aged in modern society.

In a later article, Simmons made the comparison explicit. "In anthropological perspective," he wrote, "it is literally true that societies achieved a very good old age for a few long before there could be any substantial age at all for the many. But modern civilization has reversed the process and the problems."[18] He went on to argue that, while senility was a virtually universal phenomenon, it was an ascribed status whose timing was culturally contingent. Although old age was stunningly diverse across cultures, in all societies "a point comes in senescence where any fruitful participation is regarded as past, and the incumbent is looked upon as a definite liability." Senility was not an objectively measurable state of physical weakness or mental infirmity, or both, but an ascribed status of uselessness and burden, a status that "may be attained under various degrees of physical and mental debility in different societies." Although all societies evidently made some distinction between old age and senility, the senile phase of old age "has had little significance for the simple societies which were never able to sustain more than a few really old people anyway, and those under conditions in which the very helpless could not long survive." This was not true for modern society, in which this "helpless and hopeless period of life takes on paramount importance" because economic and medical progress was successfully removing the physical barriers to old age.[19] "While modern civilization has greatly progressed in the promise of longer life for larger proportions of the population," Simmons argued, "it has disrupted many of the time-tested adaptations of the aged, and perhaps even regressed in its solution of the problem of successful aging." Yet Simmons remained optimistic. The ultimate lesson for modern society was that "the basic qualities of successful aging rest . . . upon the capacity of individuals to fit well into the social framework of their own times, to win their rights to prolonged participation and recognition, and to know when they are through." In short, society had only to cease dwelling on the problems of old age and focus instead on its opportunities; society could not solve the problems of old age for the elderly, but it could allow the elderly to create meaningful roles for themselves. "It is possible that these potentialities wait to be rediscovered, developed, and refitted into our own times," Simmons concluded. "This may, indeed, be an old frontier that calls for new pioneering."[20]

This narrative was implicit in most American psychiatric writing about senile dementia in the post–World War II period, which increasingly pointed to modern social relations as the pathology at the heart of senility. When David Rothschild described the case of a 75-year-old man who was admitted to a state hospital because of "memory impairment, mild confusion and marked untidiness of toilet habits of about one year's duration," he argued that "the striking point in his history was that he had no hobbies and few outside interests, and although he had three presumably intelligent children with whom he lived, they had made no effort to get him interested in any activities after his retirement from work at the age of 69. He would sit for hours at a time just staring into space." Thus social pathology, not brain pathology, was most responsible for producing the dementia. "It is scarcely surprising that after several years of completely vacuous existence he should regress to an infantile state," Rothschild concluded. In generalizing about such cases, Rothschild made it clear that this social pathology was characteristic of modern society. "In our present social set-up, with its loosening of family ties, unsettled living conditions and fast economic pace, there are many hazards for individuals who are growing old," he wrote. "Many of these persons have not had adequate psychological preparation for their inevitable loss of flexibility, restriction of outlets, and loss of friends or relatives; they are individuals who are facing the prospect of retirement from their life-long activities with few mental assets and perhaps meagre material resources."[21]

Some psychiatrists attempted to correlate this social pathology with dementia, much as somatically oriented psychiatrists had attempted to correlate dementia with brain lesions. A number of studies demonstrated correlation between factors such as divorce or death of spouse, living alone, recent loss of job or retirement, and admission to state mental hospitals with a diagnosis of senile dementia or cerebral arteriosclerosis.[22]

Other psychiatrists argued that brain pathology was little more than a symptom of social pathology. In a 1953 article, Maurice Linden and Douglas Courtney argued that senility, as it was usually thought of, was a "social illusion." They concluded that "senility as an isolable state is largely a cultural artifact and that senile organic deterioration may be consequent on attitudinal alterations."[23] The authors acknowledged, however, that this hypothesis was difficult to prove. David C. Wilson, writing in 1955, was less circumspect, arguing that the link between social pathology and brain deterioration was simply a matter of waiting for "laboratory proof" to support what was adequately demonstrated by clinical experience— that the "pathology of senility is found not only in the tissues of the body but also in the concepts of the individual and in the attitude of society." Wilson cited the usual hallmarks of pathological social relations in old age: the breakup of the family, mandatory retirement, and isolation. "Factors that narrow the individual's life also influence the occurrence of senility," he asserted. "Lonesomeness, lack of responsibility, and a feeling of not being wanted all increase the restricted view of life which in turn leads to restricted blood flow."[24] The narrative of ironic progress, it seemed, could be discerned even within the constricted blood vessels of the aging brain.

"A LAMENTABLE SELF-PITY": SENILITY AS A PROBLEM IN THE SOCIAL ADJUSTMENT TO AGING

Despite the observation by psychiatrists since the 1930s that Alzheimer-type dementia was more prevalent in women, senility continued to be represented primarily as a problem of middle-class men.[25] In terms strikingly similar to the late-nineteenth-century stereotype of senility, women, by virtue of the domestic focus of their lives, were thought of as immune from the worst aspects of the social pathology of senility. "Women get by pretty well," asserted physician David Stonecypher Jr. in a feature he wrote in 1957 for the *New York Times Magazine*. "A woman is not deprived of her activities as much as a man, primarily because she maintains her housekeeping responsibilities. Then, too, the woman's interests as she becomes a grandmother are similar to those she had as a mother."[26] Typically, discussions of the social production of senility centered around stereotypical male figures and problems—primarily retirement from paid labor, which was the central problem of the first generation of gerontologists and was embedded in the narrative of ironic progress that framed all gerontological discourse. When women did appear in the psychiatric literature on senile dementia, the precipitating social factor was usually a disruption of domestic life, such as the loss of a husband. For example, in a 1941 article, Rothschild cited as an exemplary case an 80-year-old woman who had been well adjusted until financial difficulties forced her to sell the family home in which she had lived all her life and her husband was sent to the town infirmary while she stayed with relatives. "She became very much upset over the break-up of her home. Previously she had been very alert, but she now seemed to 'go to pieces' rapidly. She was unable to understand why she had to leave her home. . . . Her memory became poor and she grew feeble physically . . . displayed restless and noisy conduct and failed to recognize her relatives" and died within three months of admission to the hospital with severe dementia.[27] Of course, these different representations may simply reflect a difference in the experience of older men and women whom these doctors saw as patients. But since the incidence of Alzheimer-type dementia was reportedly higher among women, psychiatrists would presumably have seen more women than men as patients. If that is true, then the fact that the cases chosen by doctors as exemplary, as well as the cruder hypothetical representations in popular accounts such as Stonecypher's, were predominantly male is an indication of how much this discourse was driven by anxiety about masculinity.

Intellectual men were often represented as less susceptible to the social pathology of senility, an idea in direct contrast to the late-nineteenth-century stereotype of senility, which represented such men as prematurely spending their intellectual capital. "Men who work primarily in the realm of ideas suffer least when they retire because it is easy for them to keep in touch with their interests and to find new ideas to stimulate them," Stonecypher asserted. "Men who work primarily as manual laborers suffer most. They cannot replace the work-day they lost with similar activities; they find themselves with time on their hands in an empty world. It is the physical laborers who are most prone to develop senile psychoses when they retire."[28] But others argued that even manual laborers could develop new and sustaining interests later in life. Continued intellectual stimulation in old age as preventative medicine was a commonplace in gerontological discourse. "Most of the great old men did not live quiet and sedate lives, free from worry or excitement," Gumpert argued. "On the contrary, they were leaders and fighters. Their brains could not deteriorate because they used them hourly in creative activity."[29] Educator Wilma Donahue asserted that even the institutionalized elderly, at risk for or already suffering significant mental deterioration, could benefit from "a program of activities which stimulates older people to express themselves and which satisfies their needs for recognition, usefulness, and self-esteem." Such a program would "tend to maintain personality integration and to retard psychological aging."[30]

The most striking difference between earlier representations of senility and its representation in mid-twentieth-century gerontological discourse was the apparent amorality of the latter. As we have seen, the moral character of the elderly was the central problem for representations of senility in both the late eighteenth and late nineteenth centuries. To the gerontological persuasion, morality was no longer an issue. Senile behavior that would likely have been categorized as sin by Benjamin Rush, and moral perversity by Beard, was described in gerontological discourse as inefficient adaptations to the psychosocial challenges of aging. "Adjustment" had become the central problem of post–World War II gerontological discourse, and both the social situation and behavior of the aging person were interpreted simply as assets or liabilities in the project of maintaining happiness and avoiding senility in old age. This discourse was rooted in the commodification of the self described by Christopher Lasch, in which selfhood was reduced to a bundle of personal attributes, acquired and zealously maintained for the status they could win on the cutthroat market of late capitalist social relations. "The modern problem of old age, from this point of view, originates less in physical decline than society's intolerance of old people, its refusal to make use of their accumulated wisdom, and its attempt to relegate them to the margins of social experience."[31] Senility, according to the gerontological persuasion, was largely the product of the radical devaluing of the attribute of age.

As shown in the examples from the psychiatric literature, the social pathology of senility was typically represented in terms of blows to the aging man's status and self-esteem. This was even more apparent in popular representations of senility. In his *New York Times* article, Stonecypher argued that, though other factors may contribute to the appearance of senility, "the *fear* and *frustration* associated with growing old in our society" were themselves "adequate to explain the whole picture of senility."[32]

These fears and frustrations fell into three groups: dread of deterioration widely associated with aging, stresses brought on by retirement, and the psychic blows incurred by the increasingly frequent deaths of friends and loved ones. But in Stonecypher's representation, all were clearly bound up with the loss of status and self-esteem. Fears of physical and mental deterioration fueled deep-seated anxieties about adequacy. "All of our lives we carry deep-seated feelings of inferiority and helplessness," he wrote. "Imagine the effect of stimulating these feelings with the belief, 'Every day, in every way, I am getting worse and worse.'" The blows retirement brought to status and self-esteem were numerous. Loss of income curtailed consumption. "The aging one faces the dismal outlook of a time in which he may be unable to afford nice clothing, the nicer foods, the newer home conveniences. The pinch hurts more when he discovers that his reduced income means that part of his independence is lost. Moreover, there is the danger of one day facing poverty." In addition, retirement was a blow to the self-confidence in his own skill that work provided. A man might be "retired from foreman at the plant to become errand boy for his wife." Finally, the death of his spouse and the increasingly frequent deaths of friends brought the loss of "a basic source of reassurance." Given the severity of these various insults to status and self-esteem, the symptoms of senility could be readily explained without even considering physical pathology. "The forgetfulness so common in senility is the senile mind's way of protecting itself from painful memories. The job he retired from, the vacations he can no longer afford, the sexual drive he fears he may lose, the state of his health, the death of his friends—these and many things associated with them may arouse thoughts intensely painful to the aging person."[33]

All the bizarre behaviors so frequently noted in cognitively impaired older people could be explained as "attempts to cope with the frustrations which our culture accords old age," Stonecypher asserted. "Jealous criticism directed at young people or, simply, general orneriness are . . . angry reactions to the deprivations accorded to old age." The persecutory delusions so commonly found were interpreted the same way. "As the patient becomes more senile he may complain that his minor body aches stem from beatings he is getting during the nights. Or he may declare the neighbors are putting poison in his food . . . [and] patients receiving excellent care may complain of neglect." Increasingly sloppy habits of personal care were seen as evidence of blows to the person's self-esteem. "The mussed hair, dirty clothing, generally unkempt appearance of the senile reveal how painful it is to the older person to care for a body which he now regards as inferior. Accident proneness can also be explained as stemming from the loss of concern for his body." Miserly behavior in the elderly was explained as an irrational attempt to conserve resources against the social onslaughts the aging man faced. "It is not uncommon for a nurse in a hospital to discover an elderly patient hoarding bits of toast or pieces of broken glass. To the desperate patient who fears his powers are slipping, any resource, even bits of toast, seems better than none at all."[34]

Though less florid in his description of senile behavior, Martin Gumpert made much the same case, arguing that the enormous emotional burden that accompanies old age would almost inevitably lead to neurosis and, if not addressed, more serious mental disturbances—which explained, according to Gumpert, the well-documented proliferation of older people with dementia in state mental hospitals. "Eliminated from participation in normal social life by the younger generation, they are often condemned without trial, without having been given the slightest chance of improvement." Gumpert thought that the feeling of inferiority so often found in the elderly was primarily caused by the loss of physical fitness, by the deformation of the body, and by their growing isolation. "But this condition is irresponsibly aggravated by a malevolent environment," he asserted. Worst of all, senility was a self-perpetuating spiral of deterioration. As the aged suffered grave personal losses, which were in turn aggravated by social pathology, their behavior all too often became so repulsive that it brought additional social rejection. "A sense of helplessness, a conviction that they are utterly useless grows in the aged, until they acquire an attitude of suspicion and hostility that merely masks their despair. They develop a tragic kind of stubbornness, a lamentable self-pity and egotism, which makes them generally disliked."[35]

Gumpert's discussion makes the problem of senile behavior clear: it was not that senile behavior constituted a breach of morality that threatened society, as Beard and other commentators of an earlier generation had insisted, but that it contributed to the unpopularity of the elderly and so exacerbated the problem of aging. Missing from all the representations of senility in gerontological discourse is the sense that the behavior of old people with senility made any difference beyond its negative effect on their social status. Psychologist Robert J. Havighurst made this point explicitly in his more formal discussion of adaptation to old age. Havighurst described senile behavior as "some of the irrational but natural defenses that a self may muster to meet the insults of old age. They are not morally wrong, but they are not particularly useful."[36]

The gerontological persuasion thus rendered meaningless all the actions of the senile man; his behavior was merely a reflection of the devaluation of old age in modern society. And if the actions of the senile man could no longer attain moral significance, neither could the actions of the majority of older people whose mental faculties were preserved. Havighurst described these as "the rational, or positive, defenses, which actually make for a happier and more highly approved life."[37] In gerontological discourse, the value of the aging man's every action was calculated in terms of his ability to make a "good adjustment" to aging. The purpose of such an adjustment, beyond the narrow goal of providing happiness for the individual, was never discernible.

The gerontological persuasion was not without a moral issue, of course. As we have seen, the project of eradicating senility and forging a new old age of health and happiness was itself imbued with great moral fervor. Gerontological discourse exuded a missionary zeal in its demand that society reaffirm the worth of the elderly. But it also displayed, sometimes in the same sentence, an intolerance of infirmity in the elderly. Thus, Martin Gumpert, in a paragraph on medical progress in eradicating senility, concluded that "senility need not be, and its presence will, in the future, be evidence of individual or social neglect."[38] In Gumpert's texts, the assertion that sickness in old age is the product of individual neglect rises at times to the level of a moral condemnation. "Most sickness is introduced into our bodies by ourselves, by our ignorance, our weaknesses and our sins," Gumpert argued. "People who suffer have to free themselves by their own power or they will never be free. Old people are neither morons nor infants. . . . Only the old man who grows to like being old, who knows no regret, who declines to make himself the object of endless self-pity, will succeed in being a 'modern' old man."[39] The theme of personal responsibility for illness was sounded more stridently in an article by a nongerontologist, prominent astronomer Harlow Shapley, who nonetheless argued that premature senility should be included on a short list of social problems that ought to be attacked with a moral fervor equivalent of that of war. After outlining the tragic waste of lives in their prime at the hands of chronic diseases associated with aging, Shapley noted that he "looked forward to the time, perhaps in a century or so, when an adult caught with a communicable disease will be heavily fined, and one indulging in afflictions like cancer, tuberculosis, arthritis, and neuroses will be branded as a social pariah, and put in jail."[40] If rhetorically less extreme, Havighurst's work on adjustment made the same point about personal responsibility. Although he had asserted that the irrational defenses that constitute senile behavior were not morally wrong and thought that the community ought to be responsible for creating conditions that allowed the elderly to create satisfying lives for themselves, he found the "conclusion inescapable that the individual, if he is to be reasonably happy and secure in his later years, must himself find rational and practicable ways of meeting his needs. No one else can do it for him. One's old age is what one makes it."[41]

In gerontological discourse, the only meaningful role left to the elderly, the one task for which they could be held responsible, was successful adjustment to old age—the creation of a personally fulfilling life. Threats to that adjustment, in the form of harmful social attitudes *or* the mental and physical deterioration of the aged themselves, could not be tolerated by the gerontological persuasion.

"AS IF THEY STILL COUNTED": RETIREMENT, RECREATION, AND THE PROBLEM OF STIGMA

If the biggest barrier to successful adjustment to aging was retirement and the loss of the meaningful role provided by work, the solution clearly lay in creating new meanings for leisure and recreation. This was explicitly argued by Havighurst and his colleague Eugene Friedmann in an influential study titled *The Meaning of Work and Retirement*. "The task set for modern man by the shift from a work-centered society to an economy of abundance with increasing leisure is that of *learning the arts of leisure*. By learning these arts well, people can enjoy retirement more than they enjoyed work," they asserted.[42]

The study, which compared the meanings and value reported by workers in six different occupations ranging from blue collar to professional, provided sociological evidence for the commonplace assumption that work meant more to men than simply earning a living. It found that workers in all the occupations (though to a lesser degree among the blue-collar occupations) reported "extra-economic" meanings for their work and that the more meaningful they found their work the less willing they were to retire. But, the authors argued, these "extra-economic meanings of work can nearly all be discovered and realized more fully in leisure activities." Work and play were essentially equivalent, they claimed. "In our economy of abundance, where work is reduced in quantity and burdensomeness to a level where it is not physically unpleasant, many of the values of play can be achieved through work and of work through play." Play offered the additional advantage of being ostensibly nonobligatory, thus offering retirees the possibility of combining freedom with the satisfactions formerly found in work. "If this secret is discovered by people," the authors concluded, "the problem of retirement will be solved, provided economic security is also generally achieved. That is, retirement from work will simply be the signal to increase and adapt one's play or leisure-time activity so as to get the satisfactions from play that were formerly obtained from work."[43]

Although the individual bore primary responsibility for adjusting to retirement, society could make it much easier by providing more facilities and programs to help the elderly cultivate the arts of leisure.[44] Failure to do so, it was frequently stressed, would intensify the social costs of senility. "With the number of people who are over 65 increasing significantly each year, our society is today finding itself faced with the problem of keeping a large share of its population from joining the living dead—those whose minds are allowed to die before their bodies do," argued recreation advocate Jerome Kaplan. "A most important preventative measure against decay of the mind is the creation of opportunities for older people to use their faculties and abilities." While such social interventions might seem expensive, they were a bargain when one considered the cost of housing the elderly in mental hospitals and other institutions.[45] In a book on the responsibility of churches to provide social opportunities for the elderly, George Gleason compared them to juvenile delinquents: "While there is little tendency for [the isolated elderly] to engage in lawbreaking operations, as juveniles do, their depredations take another form. They tend to fill our general and mental hospitals, and become a burden upon society."[46] James Woods, urging that a network of golden age centers and programs be established nationwide on the model of those in Cleveland, argued that "when we cast off and neglect the aged . . . we are creating a horde of bitter, neglected, discontented people. It is time we put something in the place of the things we take away from them." Financial support for old age was necessary but not enough, these reformers asserted. "No old person should be permitted to live out his remaining days without friends, without love, without joy, without being given a chance to use his stored-up talents. This is a task to which a recreation program addresses itself." The rationale of the drive for establishing recreation programs for the elderly was summed up by a social worker who, in ten years of work for the Public Employment Service in Los Angeles, was struck by the unhappiness of many elderly people and decided to set up a program in her church: "Older people were being pushed out of the usual organized church and community groups and relegated to the side lines. By forming the Golden Age Club we make them feel as if they still count."[47]

The language these recreation advocates used in making their case revealed the more problematic aspects of the gerontological persuasion. When Woods insisted that "no older person should be permitted" to live in social isolation, he unintentionally raised the issue of coercion—of whether the relatively powerful professionals running recreation programs were inflicting their ideas about proper ways of aging onto emotionally vulnerable individuals. Similarly, when the social worker quoted by Gleason boasts of making old people feel "*as if* they still count," she unintentionally raises the question of the authenticity of these activities—of whether such clubs and programs, even if well intentioned and pleasant, could be regarded as anything but a therapeutic contrivance that, whatever good they might accomplish, would ultimately reinforce the stigmatization of aging as a process of deterioration and increasing dependency.

These issues were central to a scathing editorial condemning recreation reformers by Edward A. Connell, the park superintendent of Stamford, Connecticut. Connell thought it bad enough that mandatory retirement policy separated so many healthy men from their office, shop, store, or factory; but what was worse was that these retirees were then "dragooned into 'Golden Age' or 'Sunset' or 'Leisure Time' clubs—virtual anterooms to cemeteries and mortuaries"—where they were "mentally dehydrated, wrung out and hung up to dry before they realize[d] exactly what [was] happening to them." Connell described the audience for a speech he gave at the local Leisure Time Club. "I winced when I saw the general physical sharpness and good health of the audience. And I thought I detected an expression of complete frustration on the faces of the men who gather once a week to listen to some speaker dilate on the glories of retirement." Connell contrasted the hollow life of these retirees, who drifted aimlessly "from checker game to movies to sidewalk superintending," with the meaningfulness of real work. His advice to the man "being forcibly retired from his job at 65 is to shun the Leisure Time and Sunset clubs like the plague and to set about getting some kind of employment." The job he found might be considerably below the one he left, but any sort of job was preferable to the stupefying life of endless leisure. He might even find it difficult to find any kind of job. "He may get boiling mad when he has been turned down time after time," Connell concluded. "But even that boiling-mad feeling is far better for him than sitting supinely in a rocking chair in some Leisure Time club while nice ladies pour tea and serve cookies—a truly poisonous diet."[48]

While this is a clearly hyperbolic account of gerontological recreation, professional discourse grappled with the same problems. What is most interesting about this discourse, and the broader discourse of the gerontological persuasion of which it was a part, is that from the outset, and with consistency, these troubling issues were explicitly addressed—yet could never seem to be laid to rest.

Regarding the issue of coercion, admonitions such as "recreation is not something you do to people or for people but *with* people" were constant in the literature.[49] But the importance of involving the elderly in the sorts of recreational activities that could prevent the slide into senility made strict adherence to this rule problematic. Thus, although James H. Woods, director of the Recreation Project for Older People of the Welfare Federation of Cleveland, began his discussion of the issue by saying that the professional leader of a program would have little need to assert authority over any of the elderly participants, he went on to cite the opinion of the head of a state mental hospital that "many old people retrogress, not because it is an inevitable result of aging, but because they sit and do nothing." This psychiatrist went on to assert, according to Woods, that "any procedure, even the use of authority, was justified to get these people out of their ruts." While authority even in extreme cases had its limits, Woods felt that if "the activity would be beneficial for some old person, [the recreation professional] might feel justified in attempting to persuade as strongly as he knows how in the interests of good mental health."[50] Carol Lucas, an educator at Columbia Teaching College who trained recreation professionals, approached the problem much the same way. "Recreation activities cannot be forced upon older people, cannot be superimposed simply because it is good for them. If the activity is truly beneficial in effect, then it must be chosen, it must be wanted by the person participating," she began. But she went on to argue that this did not mean "that the director or leader cannot tempt, subtly persuade, or gently entice an elderly person into an activity for which he has little enthusiasm—and with highly successful results. Each director tailors his or her technique of persuasion to the individual involved." Moreover, it was the director's obligation to disabuse their elderly clients of their misplaced attachment to work. "It is necessary to interpret constantly to the senior citizens with whom we work the importance of recreation, to make it clear to them that play is just as respectable and dignified as work," Lucas wrote. "Don't forget, they have spent the greater part of their lives in a work-centered society that looked on recreation as an indulgence or an activity only for children."[51] In other words, while the professional should never force elderly clients into a program they did not want, the professional was equally obliged to persuade elderly clients to want to participate in those programs.

The tension so clearly evident in gerontological discourse between respect for the autonomy of the elderly client and the need for therapeutic intervention does not mean that professional leaders of programs for the elderly were simply agents of social control, disciplining their aging subjects to conform to the life-course pattern demanded of them by consumer culture. Such a conclusion remains possible, of course, but would require the type of evidence mustered by a thorough social history of professional-client relations—far beyond the scope of this project. But paying attention to this tension in gerontological discourse does allow us to ask a more fundamental question: Is there a meaningful distinction to be made between coercion and therapeutic intervention?

A case briefly reported in 1959 in the journal *Recreation* is instructive. Mr. M., a wheelchair-bound, permanent resident at the Milwaukee Veterans Administration Hospital, was "an eighty-two-year-old Spanish War veteran, . . . a paraplegic spastic, with blindness of the left eye, and generalized arteriosclerosis." At an initial interview, the hospital's special services coordinator (and author of the article in which this case history appeared) learned that in the past Mr. M. had had an active interest in playing folk music on the violin. "He showed us his violin quite proudly, but informed us that his fiddling days were over. We asked him to play a tune but he refused." But, the superintendent noted, Mr. M. "apparently reckoned without our persistence because, whenever we encountered him thereafter, we would ask him when he was going to play for us. Finally one morning, obviously weary of our persistent queries, he said 'All right, if you want music, be in your office Monday morning at nine o'clock and I will bring some real musicians to play for you.'" At the appointed time, Mr. M. arrived with two musician friends and the three held an hour-long jam session in a small recreation area at the hospital where a crowd soon gathered. At the end of the program, the staff asked if the trio would be willing to come back on a weekly basis, to which they readily agreed, bringing a variety of other musicians and singers with them; the jam sessions soon became one of the most popular events at the hospital, and Mr. M. derived a great deal of satisfaction from his role in leading them.[52]

Obviously, we would like to have more than the superintendent's account of this case before drawing solid conclusions, but on the face of it the professional intervention can only be regarded as coercive—the continual badgering of a patient about an activity in which he had clearly declined to participate until the patient eventually relented. At the same time, unless the author is simply lying, we must also regard it as a highly successful intervention—restoring a significant measure of self-worth to a man who had apparently given up on life. The coercion exercised by the superintendent was not justified by the success of the intervention but by Mr. M's need for that intervention, and his need was shared in greater or lesser degree by *all* aging people. Thus, it was not that the professionals running golden age programs exercised coercive power to "dragoon" hapless retirees into participation nor that their therapeutic interventions were always benign and respectful of the autonomy of their clients. The recreation professionals had a wide range of techniques available, ranging from unintrusive supervision to coercion to accomplish their work. The successful professional would, as Lucas taught, tailor them to fit the needs of the individual client in order to help him or her make a good adjustment to aging.

What was most problematic about the discourse on recreation and the gerontological persuasion in general was that it brought every aspect of the life of every aging individual within the realm of professional scrutiny, into what Stephen Katz has called the "gerontological web."[53] Even an aging person who showed no signs of mental deterioration, who participated—or didn't—in various organized recreation activities according to his or her own wishes, nonetheless remained implicitly under the scrutiny of professionals who stood ready to intervene, with coercion if necessary. Although the aging individual might be able to demonstrate that he or she was not yet senile, he or she could never avoid the question. If the problem of geriatrics has been the inability to differentiate between the normal and pathological in aging, gerontology has deepened the problem in the social realm by subjecting even what it passionately insisted was normal to the strictest scrutiny for signs of the pathological. Thus the gerontological persuasion, in trying to salvage old age from the junk heap to which it had been consigned by modernity, actually deepened its stigmatization.

Leaders in gerontological recreation were aware of this and struggled against it. Woods criticized the tendency among both professionals and the public to treat "older men and women as somewhat in the nature of curiosities, different from the rest of human kind. When they fall in love, do a good job of work, show themselves interested in the same things that bring pleasure to the rest of the world, it is looked upon as a matter of surprise and wonderment."[54] But the very programs Woods advocated were based on ideas of difference. "A Golden Age Club is more than a collection of people," he wrote in his manual on establishing and running recreation programs. "First of all, it is a group of *old* people. That fact is the element of difference. While individual old people may display a remarkable state of physical preservation and fitness, they are the exceptions. Disabilities must be taken into account when the program is being planned." Not that this difference was to be regarded as the essential characteristic of older people, for "the second thing to remember about the Golden Age Club is that in addition to being old, its members are *people*. Because they are people they have fears, hopes, frustrations, desires—all the great orchestra of feelings that play within the human soul."[55]

So vexing was this problem of difference that Woods devoted a separate chapter to it: "Are Old People 'Different'?" For younger people who might be considering entering the field, Woods sought to allay their fears that they would be dealing with people who would be difficult to understand or please—"a mass of crabbed individuals who dislike new ideas, loud noises, drafts, and the ways of the younger generation generally." Woods began with the typical assertion of the similarity of older people to those of other age groups, that "there is as much variety of personality, habits, likes, and dislikes among the old as in any other group." Nonetheless, in a large group of older people, certain peculiar characteristics—"part not necessarily of the aging process but rather of the total life experience and the general insecurity of older people in our society"—would appear with greater frequency than in a group of younger people. It is to these special characteristics that the bulk of the chapter is devoted, characteristics such as "feelings of inferiority, inadequacy, and frustration . . . mental deterioration . . . loneliness . . . death wishes . . . [problems associated with] money . . . overconcern with bodily ailments . . . [and] conservatism, a preference for the tried and familiar ways of the past."[56] Thus, in an avowedly optimistic book demanding the establishment of programs to help older people enjoy life, we find a reiteration of virtually all the negative traits ascribed to them.

Such was the stigmatizing logic inherent in the gerontological persuasion. This is not to deny that the programs and policies implemented by gerontological activists to improve the lives of the elderly accomplished much good or that the claims of leaders in gerontological recreation that golden age clubs and programs were tremendously popular among the elderly themselves are to be disregarded. But it does explain why, despite the insistent optimism of post–World War II gerontology, senility continued to cast a gloomy shadow over the golden age.

FIGHTING THE "MYTH OF SENILITY": THE POLITICIZATION OF THE GERONTOLOGICAL PERSUASION

Beginning in the late 1960s, the gerontological persuasion took on a more stridently activist tone, with many of its leaders consciously following the language and methods of the apparently successful civil rights and women's liberation movements. "Ageism" became one of the central concepts of the gerontological persuasion. The term was coined in 1968 by Robert N. Butler, who became perhaps the most visible proponent of the gerontological persuasion after winning a Pulitzer Prize for his book *Why Survive? Being Old in America*. According to Butler, "ageism can be seen as a process of systematic stereotyping of and discrimination against people because they are old, just as racism and sexism accomplish this with skin color and gender. Old people are categorized as senile, rigid in thought and manner, old fashioned in morality and skills. . . . Ageism allows the younger generation to see older people as different from themselves; thus they subtly cease to identify with their elders as human beings."[57] Butler saw ageism embodied in myriad ways: in stereotypes and myths; disdain and dislike; subtle avoidance of contact; demeaning jokes and representations in the mass media; discriminatory practices in housing, employment, and services of all kinds. It operated on both the institutional and individual level, at once functioning to relieve society of responsibility for the aged and protecting individuals from confronting fears about their own aging. Perhaps worst of all, ageism was internalized by the elderly so that they tended to perpetuate the very stereotypes directed against them.[58]

The parallel between the elderly and racial and ethnic minorities had first been struck in the 1950s. The analysis by social gerontologists of the way in which the problems of the elderly were created by negative attitudes and discrimination led logically to a comparison with racial and ethnic minorities, whose suffering from a hostile social environment was long-standing and broadly recognized—at least in academic circles.

In 1953 sociologist Milton Barron proposed the "quasi-minority group" status of the aged as a theoretical framework for study of the aged.[59] Like racial and ethnic minority groups, Barron and those who followed him argued, the aged were marked as visibly different, were subject to categorical prejudice and discrimination—for example, through mandatory retirement and informal mechanisms denying them access to employment—and were the object of negative stereotypes and images that circulated at every level of society. In response, the elderly were manifesting both the positive and negative signs of a minority group subculture. On the one hand, the elderly were internalizing much of the social hostility—which, in fact, explained many of the emotional and behavioral problems associated with old age and, somewhat paradoxically, also explained why many older people were unwilling to identify with or acknowledge their own aging. (As late as 1980, a gerontology monograph cited Stanley Elkins' argument that slavery produced a defective personality type—the Sambo—and argued that the social pathology around old age acted as a similar mechanism to produce much of the mental deterioration labeled "senility.")[60] On the other hand, "for the growing minority that has reacted against the negative self-conception characteristic of aging in our society and has seen the problems of aging in a group context . . . there are expressions of group pride . . . and a self-acceptance as a member of an esteemed group and the showing off of prowess as an elderly person (for example, in 'life begins at 80' types of activities)."[61]

The minority group theory of aging was controversial from the outset. In a widely cited article, Gordon Streib argued in 1965 that the notion that the aged constituted a minority group obscured more than it revealed. The aged did not share a distinct and separate culture but retained long-standing identification with other groupings. Moreover, membership in the "group" was not exclusive and permanent but awaited all members of society who live long enough. Thus "age is a less distinguishing group characteristic than others such as sex, occupation, social class, and the like."[62] The theory was defended but remained controversial.[63] But the broader point—that the worst of the elderly's problems were not caused by aging per se but by the hostile attitudes, irrational fears, and barriers to full participation in life imposed by society—remained the conventional wisdom of the gerontological persuasion.

Thus Butler's concept of ageism was very much like what gerontological advocates had been saying for decades. What *was* new was its explicit comparison to racial and sexual prejudice, a link that logically implied the sort of direct political action that had been successfully used to win progress for those groups. "The elderly are becoming more aware of their fate and more sophisticated about political action, and it has been instructive for them to watch the protests of the civil rights, peace and women's movements of the 1960s and 70s," Butler wrote. This nascent activism would grow, he predicted, as the elderly increasingly reached retirement in good health and constituted a larger proportion of the total population.[64] Where earlier gerontological advocates had framed the problem of old age as one of both personal adjustment for each elderly individual and the professional administration of social interventions, Butler conceived of it largely as "the politics of building an appropriate and dignified old age for Americans."[65] Channeling the pent-up frustration and emotion of the elderly themselves into political action was perhaps the most important strategy of the post-1960s gerontological persuasion, and Butler provided a detailed "Agenda for Activism" along with information on the various organizations for older people that had emerged in the 1970s.

Although political interest groups for the aging such as the American Association of Retired Persons, the National Alliance of Senior Citizens, and the National Council of Senior Citizens were perhaps more powerful—boasting many millions in membership—the activist tone of the gerontological persuasion was perhaps best epitomized by the emergence of the Gray Panthers, organized in 1970 by Maggie Kuhn after she was forced to retire at the age of 65 from her job with the Presbyterian Church. The Gray Panthers remained a relatively small (about 40,000 members by the 1990s) though high-profile organization that differed from other aging interest groups in its cohesiveness around left-liberal political values. Kuhn saw the Gray Panthers as part of a larger movement of aging people that merged with a broader, intergenerational agenda for social change. "Many of us have realized we have reached our seniority in a new age—an age of liberation and self-determination," she wrote in 1976. Some older people were uncomfortable in this new age, she allowed, and resentful of the changes around them, or they were resigned to a society that has no place for the elderly or were filled with self-hatred and hostility toward their aging selves. But "an increasing number of us are outraged about the injustice and dehumanization we see and are ready to work for basic societal change."[66]

Another popular book of the mid-1970s, Alex Comfort's *A Good Age*, also openly acknowledged the debt owed to the civil rights movement. The book consisted of an introduction that functioned as a sort of manifesto for the politicized gerontological persuasion, followed by an encyclopedia of aging, which Comfort characterized as "facts, self defense measures, problems and also some of the plus factors and pleasures of later life" with which the elderly could "de-brainwash" themselves. Interspersed with the entries were scores of brief biographies, with elegant illustrations, of seniors who continued living productive lives well into old age. The biographies ran the gamut from John Wayne to Gandhi, from world-famous celebrities to common men and women, and were obviously meant to illustrate both the diversity of successful aging and the shared resistance to demeaning stereotypes of deterioration—a veritable catalog of positive images of the elderly.

Repeatedly, Comfort invoked blacks' protest of racist stereotypes as the model the elderly would have to emulate. "React to people who talk slightingly about seniors ('old duffer,' 'old biddy,' 'dirty old man,' 'old lady in tennis shoes') in the way that black people have learned to react to people who talk slightingly about 'niggers,'" he exhorted. "Tell them you don't appreciate that sort of language. Your reaction will give them a salutary shock. Usually they mean no harm, but need their heads changed, to see older people as people, and only secondarily as old. . . . This is part of the shaping of society."[67] Seniors would also have to protest demeaning images in the mass media as blacks had done, especially since they were becoming a growing segment of the audience. "There is a means here of catering to needs, of giving information and of stereotype busting, which will need a sharp effort like that which got black figures more dignified and realistic treatment in entertainment. Short of this, the scripts will go on falsifying age along the old lines."[68]

As with the liberal attack on racism, the logic of the attack on ageism required that the elderly could be shown to be essentially no different from other age groups. This tendency regarding race can be seen in historian Kenneth Stampp's now notorious claim that "Negroes *are*, after all, only white men with black skins, nothing more, nothing less." Stampp did not deny that there were cultural differences between whites and blacks; nor did he mean to imply, he made clear, that Negroes should feel flattered by the comparison. Stampp simply wished to assert that there were no "significant differences between the innate emotional traits and intellectual capacities of Negroes and whites."[69] In a strikingly parallel construction, Comfort argued that within every older person "there is a young person, the same person, inside. Older people are in fact young people inhabiting old bodies and confronted with the physical problems of reduced vigor, changing appearance and, although many escape these, specific disabilities affecting such things as sight and agility." Science would make the truth of this more apparent as it developed the technical ability to "suppress or slow the physical changes." But Comfort did not wish to deny the dignity of the elderly by relying on this technical solution, "rather as if we say that we could end race prejudice if we could turn all black people white." Seniors should feel no more flattered to be compared to young people than blacks should to be compared to whites. Age had a dignity of its own.

Of course, the "race-blind" liberal position Stampp's quotation came to exemplify was attacked precisely for its tendency to "turn all black people white" by glossing the creativity and distinctiveness of Afro-American culture or, worse, construing that cultural distinctiveness in pathological terms, as in the controversial Moynihan report titled *The Negro Family*, which, in order to make a case for public assistance, depicted the African American family as embedded in a culture of poverty.[70] The liberal attack on ageism that the work of Butler and Comfort exemplifies exhibits the same difficulties. Despite its wish to assert the dignity of the elderly in their own right, it cannot shake the accusation that it simply demands old people to be young. The differences between old age and youth are pathologized. Significant physical differences are to be regarded as the product of discrete disease entities rather than of aging, while psychological or emotional differences, such as detachment and disengagement, are to be regarded as mental illnesses. "We must be skeptical of the liberal assumption . . . that age is irrelevant, that older people differ from younger people only in their chronological age," Thomas Cole argues in *The Journey of Life*. "In age, as in race and sex, the Scylla of prejudice is never far from the Charybdis of denial of human difference—differences that need to be acknowledged, respected, and cherished."[71]

With regard to senility, the tendency to regard differences in old age as discrete disease entities made the broad conceptualization of senility characteristic among post–World War II American gerontologists unacceptable. *Senility* was a waste-basket term for a variety of discrete disease entities, many of them treatable. The continued currency of the term among professionals was, according to Butler, a product of their conscious and unconscious fears of and hostilities toward old age. To Comfort, its continued use was a means of silencing uppity elders: "Going out of your head, losing your memory or becoming 'senile' is statistically an unlikely misfortune," he argued. "The word 'senile' itself is less a diagnosis than a term of abuse—you're senile if you make waves. That indicates only that your brain is still functioning and they haven't washed it for you."[72] To the politicized gerontological persuasion, the right to a real diagnosis was a fundamental right of the elderly. As we shall see in chapter 5, it was a key element of the Alzheimer's disease movement that emerged in the 1980s.

CHAPTER FOUR

The Renaissance of Pathology

In the late 1960s, a growing coterie of psychiatrists and neurologists began to take a fresh look at the problem of dementia in old age. To this group of researchers, the broad concept of "senility" that had been developed a generation earlier by psychodynamic psychiatrists and embraced by gerontologists and other professionals concerned with aging in the 1950s and 1960s seemed archaic at best, a throwback to the psychoanalytic hegemony that was fast receding in the wake of new techniques being developed in the neurosciences to study the structure and function of the human brain. Moreover, many of these researchers shared the attitude of the gerontological persuasion and agreed with Robert Butler and others that the very term *senility* was more an ageist epithet than a meaningful medical concept.

In the new world of Alzheimer's research that this group created, brain pathology became the essential feature of senile dementia, the symbol of its grave reality and the object of cutting-edge biomedical research. In part, this was the result of technical and conceptual developments in various aspects of neuropathological research—more rigorous correlations of brain pathology and clinical dementia, electron microscopic studies of the senile plaques and neurofibrillary tangles, and the measurement of neurotransmitter levels in the diseased brain. But on a more fundamental level, meeting the demand that senility be taken seriously involved reframing it as a scientific rather than a social problem. Of central importance was the relationship of Alzheimer's disease to normal aging. Research had to be about more than the processes of aging; it had to be about something real and immediate—a dread disease. The broad concept of senility that had been constructed through the psychodynamic model of dementia had to be broken back down into its constituent parts. Thus Alzheimer's disease had to be conceptualized as an entity distinct from the aging process. To do this, scientists employed analogies between senile dementia and other credible diseases and used metaphors that bridged the gaps between theory and data in a way that cast mental deterioration in old age as a problem suitable for cutting-edge medical science. Ultimately, Alzheimer's disease had to be formulated in a way that was consistent with the central tenets of the gerontological persuasion—that deterioration and disability in old age were the result not of age but of disease.

"REFLECTED IN THE ULTIMATE APPEARANCE OF THE BRAIN": THE CORRELATION OF PATHOLOGY AND DEMENTIA

In the 1960s, the most immediate barrier to making senile dementia a problem for cutting-edge medical science was the apparent insignificance of its pathological hallmarks. As discussed in chapter 2, Rothschild and his followers had used the failure to establish a correlation between the degree of pathology found at autopsy and the presence and degree of dementia in life to argue that psychological and social factors, not the plaques and tangles found in the brain, were the major causal factors in dementia, or at least the factors most worth studying because they suggested possibilities for research and treatment. Theorizing dementia this way also made it a problem that could be investigated by psychodynamic psychiatry, which from the 1940s to the 1950s was, in America at least, widely regarded as the cutting-edge science of the mind. But by the 1960s, at the dawn of the biological revolution in psychiatry and the emergence of the interdisciplinary field of the neurosciences, this approach seemed increasingly antiquated to a growing number of researchers, and the connection between pathology and the clinical manifestation of dementia required a reexamination.

The first significant step in this direction was the efforts of researchers in Britain to establish firm correlations between pathology and dementia through two large-scale research programs that combined the traditional histopathological findings of light microscopy with statistical methods. It was no surprise, given the resistance of British psychiatry to the psychodynamic model of dementia, that the renaissance of pathology would begin there. The most prominent figure in this work, Sir Martin Roth, had been a harsh critic of Rothschild's work in the 1950s (see chapter 2). But the first large-scale effort to reassert the importance of pathology was undertaken by J. A. N. Corsellis, a neuropathologist at Runwell Hospital in Essex. Corsellis argued that the influential work of Grunthal, Gellerstedt, Rothschild, and others, which was the basis for the assertion that pathology was not significantly correlated with dementia, should not be used to make broad generalizations because they compared intensity of clinical and pathological findings among selected groups and not the overall population. "These and similar investigations have yielded valuable information," Corsellis argued. "But none was primarily designed to interrelate the different patterns of pathological change that might be observed if an adequate sample of the whole population were systematically studied."[1] Carrying out a study that involved clinical and pathological evaluations of large numbers of older people in the community was beyond the resources available to him, but Corsellis argued that he could make the population of a mental hospital serve as a workable substitute for the community by dividing the functional from the organic disorders.[2] Since there was no great difference in the mean age at death between patients in these two divisions, Corsellis argued that a valid comparison of the degree of pathological change that developed over time in the two groups could be made.

Corsellis analyzed separately and then correlated the clinical and pathological findings of 300 consecutive elderly patients who came to autopsy at Runwell. Rating the degree of various types of pathological damage (vascular lesions, cerebral atrophy, senile plaques, and neurofibrillary tangles) on a four-point scale, he was able to analyze the data statistically. His results showed that there was a clear tendency for cerebral degeneration of these various types to increase with age to an approximately equal degree in both categories, but the organic group showed a significantly higher degree of all types of cerebral degeneration at every age. Corsellis concluded that his data suggested a "well-ordered and demonstrable pattern" for the common forms of cerebral degeneration—a finding that he thought demolished the psychodynamic model and restored pathology to its rightful place. "In the past it has been widely accepted that the extent to which any brain may be damaged, particularly in later life, is so variable and so unpredictable that even the broadest correlations are difficult to establish and may well not exist," Corsellis noted. This resulted, thanks mostly to the work of Rothschild, "in a swing to the other extreme . . . and cerebral pathology was relegated, if not to a back seat, at least to a long bench on which many other equally important factors were immediately installed." Corsellis conceded that Rothschild had made a "great contribution" in drawing attention to "the possible influence of other, non-pathological factors, on the way in which an individual may be altered clinically by progressive degeneration." But Rothschild and those who followed him had made too much of "anomalous instances," prematurely dismissing pathology without addressing the question of "how often the brain might appear to function efficiently while it disintegrated, or conversely how often the patient with clinical evidence of demen-tia might turn out to have an apparently intact brain." The answer, Corsellis concluded, was clear: "the ways in which 'organic' deterioration manifests itself during life [i.e., dementia] are more often than not reflected in the ultimate appearance of the brain." In the overwhelming majority of cases, dementia was associated with pathological changes in the brain.[3]

The second major program to investigate the issue of clinical-pathological correlations was organized by Martin Roth, who had argued in the 1950s that much of the evidence amassed by American psychiatrists in support of the psy-chodynamic model of dementia was in fact the product of nosological confusion and diagnostic sloppiness (see chapter 2). In the town of Newcastle upon Tyne, Roth led a Medical Research Council group on the relation between functional and organic psychiatric illness, which also included pathologist Bernard Tomlin-son and psychiatrist Gary Blessed. The Newcastle study, like that of Corsellis, was specifically aimed at Rothschild's claim that pathology was not necessarily the most important cause of dementia.[4] Although the Newcastle researchers studied only seventy-eight cases (fifty patients with dementia from the mental hospital and, as controls, twenty-eight patients without dementia from the mental hospital and the geriatric ward of a general hospital), a number of important refinements in their methodology allowed them to present a case for the significance of pathol-ogy which was even more persuasive than that of Corsellis. First, it was a prospec-tive study—assessing patients in life and following them to autopsy rather than beginning with an autopsy and retrospectively studying their clinical histories. Second, it included as controls patients who were assessed as normal with psy-chological exams specifically designed to reveal dementia. Most important, unlike Corsellis, who had simply grouped his case material into all-or-none categories and used arbitrary categories to describe the degree of pathology—that is, "none, slight, moderate, severe"—the Newcastle researchers developed procedures for quantifying both their clinical and pathological descriptions. Blessed employed standard psychological tests and a quantitative dementia scale he developed, and Tomlinson established a standard method for counting plaques that could be verified by others. (Neurofibrillary tangles, vascular lesions, cerebral atrophy, and other pathological structures associated with dementia were not amenable to such techniques and so had to be assessed in more subjective terms, as Corsellis had done.) With both the clinical and pathological findings quantified, the re-searchers could calculate correlation coefficients to describe the significance of the associations they observed. Moreover, if subsequent researchers employed the same procedures, they could generate comparable data sets.[5]

In a series of heavily cited articles appearing from 1966 to 1970, the Newcastle group reported its results. In the researchers' initial article in *Nature,* they proclaimed that "far from plaques being irrelevant for the pathology of old age mental disorder, the density of plaque formation in the brain proves to be highly correlated with quantitative measures of intellectual and personality deterioration."[6] Ultimately, the Newcastle group found that the correlation coefficient was .77 for the association between plaque counts and Blessed's dementia scale and .59 for the association with psychological tests—both results being highly significant statistically.[7] Although other forms of brain pathology could not be quantified as rigorously, they too appeared to be strongly associated with the degree of dementia reflected in the clinical findings. Age, by contrast, was not significantly associated with dementia scores (.13). Even though the difference between the two groups was purely quantitative—every pathological feature of the patients with dementia was also found to a lesser degree in the patients without dementia—it was a clear and significant difference. Seventy percent of the cases showed more pathological change than any control, and severe pathological deterioration of all types was restricted almost entirely to the group with dementia. The authors estimated that fully 90 percent of the cases in the group with dementia could be satisfactorily accounted for by pathological changes that distinguished them from the control group. That 10 percent of the cases of dementia could not be explained pathologically was not surprising to the authors, for perfect correlation "would mean that the pathological associations of all dementias in old age are recognized." Occasional anomalies no doubt occurred, but the authors concluded that there was no evidence to support Rothschild's claim that equally severe pathological destruction could be found in normal old age as found in senile or arteriosclerotic dementia.[8]

The Newcastle study did not resolve all issues regarding the relative importance of pathology, however. First of all, anomalous cases were indeed present in the researchers' data. For example, six subjects who had very low dementia scale scores were found to have moderately high plaque counts. To account for this, the authors had to employ a weak version of Rothschild's model, acknowledging that individuals might have differing ability to sustain pathology but restricting that differing ability to the "reserve capacity of the cerebrum."[9]

More important, the correlations themselves were not without ambiguity. It was possible, the authors acknowledged, that the very high correlations were caused mainly by the presence of the controls, who had very high correlations between pathology and dementia, that is, very low plaque counts and no dementia. When the group with dementia was considered alone, the correlations were in fact much less impressive—.4 for the dementia scale (weak statistical significance) and .26 for psychological tests (no statistical significance). The authors suggested that this might be because, beyond a certain threshold point, the degenerative process measured by plaque counts has done maximum damage to cognitive functioning, raising the possibility that people with severe dementia are separated from the normal elderly by some as yet undefined qualitative factor.[10] But it was also possible to conclude, as did Tom Kitwood, an advocate in the 1990s for a return to the psychodynamic model, that plaques and tangles are not the only factors worthy of study in trying to determine the etiology of dementia.[11]

The Newcastle study also had an inherent limitation, namely, that statistical correlations could not necessarily be used to infer causation, and though the authors indulged in etiological theorization at some points, they were always careful to indicate that these were highly speculative and untestable by the evidence at hand. Thus the Newcastle study itself did not lead to any significant changes in etiological theory or shed any new light on other long-standing conundrums regarding dementia in late life. For example, the study provided no new evidence regarding the long-standing debate about whether senile dementia or Alzheimer's disease (or both) should be regarded as a disease or an exaggerated form of the changes associated with aging. As mentioned, the study showed a quantitative rather than a qualitative difference between normal aging and dementia. Thus, it remained reasonable to argue that people with dementia were simply more "aged" cerebrally, that they were simply further along in a process that affects every aging individual. Although the authors speculated that an as yet undetected qualitative difference might exist, produced by some as yet unknown pathological process, the study offered no significant evidence one way or another. The study also confirmed, without being able to explain in any way, a puzzling observation that had been noted anecdotally since a paper by Rothshild in 1934.[12] There was a significant difference in the distribution of dementia by sex, with women being more likely to develop Alzheimer-type dementia than men, whereas the opposite was true for arteriosclerotic dementia. Because Alzheimer-type dementia was much more common (about 50% of the Newcastle cases) than arteriosclerotic dementia (about 17%), women overall were significantly more likely to develop dementia.[13]

Despite these problems and limitations, from the time in which these articles were published, the Newcastle study was widely regarded as providing authoritative evidence for a strong and probably causal relationship between pathology and dementia.[14] To believers in the importance of pathology, the study provided substantial, if not quite conclusive, proof. The turn toward the biological was further enhanced by the application of electron microscopy to the pathological structures that had first been described at the turn of the century by Alzheimer and his colleagues.

"LET THE BIOLOGISTS GO AS FAR AS THEY CAN": NEUROPATHOLOGY AT THE ULTRASTRUCTURAL LEVEL

The application of electron microscopy to the pathology of Alzheimer's disease by two researchers working separately, Michael Kidd in Britain and Robert Terry in the United States, opened up new and exciting ways of exploring problems of neuroanatomy that had for decades hindered research into the pathology of dementia. Electron microscopy's importance was not that it could authoritatively resolve any of the grand conceptual issues regarding the relationship of pathology to dementia, and dementia to normal aging, as has sometimes been suggested,[15] although it did quickly become the arbiter of some particular issues. Quite the opposite, electron microscopy transformed the familiar and unproblematic pathology of dementia into something new and strange. By extending the researcher's field of observation to a new "ultrastructural level," electron microscopy created a new world in which basic features of pathology had to be interpreted afresh. Pathologists entering the ultrastructural level were not completely in the dark, of course; much of their interpretive work involved relating what they knew from light microscopy to what they observed at the ultrastructural level. But the insights of traditional histopathology could only be extended so far, and the new technology forced researchers to pose new questions and formulate new theories—stretching the neuropathology of dementia toward the frontiers of molecular biology.

The most obvious illustration of how electron microscopy transformed the familiar and stable pathology of Alzheimer's disease into something that required interpretation is the disagreement in the early studies concerning the structure of the neurofibrillary tangle. Electron micrographs of neurofibrillary tangles showed that, though they contained normal neurotubules and neurofilaments, their most striking feature was thick bundles of parallel tubules, or filaments, that wound their way through and displaced normal elements of the neuron. Higher-resolution images revealed that the individual tubules or filaments narrowed at regular intervals.

Both Kidd and Terry published articles in the early 1960s which interpreted these images differently. In 1963 Kidd published a brief article in *Nature* and the next year a longer article in *Brain,* in which he argued that the structure of the abnormal material in the tangles was a paired helical filament.[16] In articles published in the same years, Terry described the structure of the neurofibrillary tangles as a twisted neurotubule—the periodic bulging and narrowing produced by constriction of the tubule at the twist.[17] The difference was important because ultrastructural features provided the best clues available to the biochemistry of pathological structures. (Electron microscopy revealed that the evidence provided by traditional histopathological methods had severe limitations.) Both interpretations were viable given the evidence, but for reasons that are not apparent in the professional literature, Terry's was the conventional description until the mid-1970s. For example, throughout the proceedings of a 1969 Ciba Foundation symposium on Alzheimer's disease, in which both Kidd and Terry were participants, the tangles were referred to as bundles of twisted tubules—except when the issue of their structure was explicitly addressed. Here, the proceedings reveal that the actual structure of the material composing the tangles remained open to question. At one point, Kidd confessed, "When I suggested that the abnormal structures in the electron microscopy of this disease were twisted, I was aware that they could be twisted tubes, but I felt that they were more like twisted pairs of filaments. Now I am not sure which they are."[18]

In 1976 Henryk Wisniewski, H. K. Narang, and Terry published a paper that resolved the issue, showing that Kidd's interpretation had in fact been correct. Using a "tilt stage" electron microscope to produce images of the "tubules" at carefully defined angles, they constructed a wire-mesh model in the form of a double helix. When cut and positioned in ways that were analogous to the electron micrographs of actual tangles, x-ray images of the double-helix model were essentially identical to the tangles seen in electron micrographs.[19]

Although the images produced by electron microscopy created new and sometimes difficult interpretive issues, they also allowed neuropathologists to address meaningfully basic questions on which no progress had been made for decades—how were the plaques and tangles related to each other, and what was their biochemical composition? At the level of light microscopy, the only clue to the biochemical composition of pathological structures was their receptivity to the dyes that were used to stain tissue sections. A shared affinity to a particular dye was taken as in indication of shared biochemical properties. Thus, the receptivity of senile plaques to Congo Red dye led Belgian researcher Paul Divry to conclude in the 1930s that the plaques were composed of amyloid because amyloid was also "congophilic." Since tangles were congophilic, Divry assumed that they were composed of some form of amyloid as well. All researchers did not agree with this, but no methods were available which could resolve the issue.[20] In his 1964 paper, Terry confirmed that the core of the senile plaque was amyloid by observing morphological features that were identical to those in other tissues proved to be composed of amyloid. Moreover, he was able to describe other features within the plaques visible only at the ultrastructural level including the presence of "twisted tubules" in neurites (the components—axons and dendrites—of the synapse) engulfed by the plaques. Despite the new finding of twisted tubules within the plaque and the tangle's affinity for Congo Red dye, Terry found that at the ultra-structural level there were morphological differences that indicated that the tangle was composed of something other than amyloid. Terry argued that this was indicative of the limitations of traditional methods. Staining properties were based on a match between the size and orientation of molecules in the dye and material, but, as he and Wisniewski put it in their contribution to the 1969 Ciba symposium on Alzheimer's disease, the staining reaction could in fact be "deceptive and non-specific."[21] In light of this and other problems, another pathologist who had worked with Terry on the study published in 1964, Nicholas Gonatas, made a "plea for some reconsideration of the criteria for identifying amyloid."[22] Reliable identification of amyloid required the observation of morphological features at the ultrastructural level that were identical to amyloid found in other tissues.

Beyond providing descriptions that could be used to infer more powerfully the biochemical properties of the plaques and tangles, electron microscopy inspired researchers such as Terry and Wisniewski to hypothesize about the genesis of the pathological structures they observed. In their Ciba symposium paper, Terry and Wisniewski hypothesized, on the basis of the presence of "twisted tubules" (paired helical filaments) in the neurites, that the deposition of amyloid and formation of plaques were a secondary and perhaps coincidental change. The tangles were the primary degenerative change—disrupting the flow of plasmic material down the axon (the arm of the neuron which "sends" a "message" through the release of neurotransmitter chemicals across the synapse to be absorbed by the dendrites of neighboring neurons).[23] In the discussion of this paper, one conference participant expressed bewilderment at this hypothesis because at the level of light microscopy it was not uncommon to find cases of plaque-only dementia. Terry responded, "I am puzzled by all sorts of imbalances here that I recognize and am not yet prepared to deal with."[24] Ultrastructural evidence produced by electron microscopy was clearly far in advance of the ability of researchers to theorize it fully.

Taken as a whole, the proceedings of the three-day Ciba Foundation symposium, seen now as a landmark in the emergence of Alzheimer's disease as a project for modern biomedical science,[25] illustrate the importance of electron microscopy in transforming Alzheimer's disease research. Although the evidence this new technology generated was perhaps difficult to organize into a theory, it was rich enough to provide neuropathologists with many interesting and potentially fruitful directions to pursue. The excitement that electron microscopy generated can be seen in the closing remarks of Martin Roth, who chaired the symposium: "The observations that have come from the ultrastructural studies . . . are impressive for their depth and variety. . . . Even certain biochemical activities have been tentatively established in association with certain structural changes. . . . Such differentiation of the pathological changes could not have been conceived even seven or eight years ago and it raises the scientific study of Alzheimer's disease to a new level."[26]

Roth clearly viewed the electron micrograph as an emblem of scientific progress and pointed to the various ways in which ultrastructural evidence presented at the symposium had supplanted traditional methods, such as its resolution of issues regarding the amyloid basis of the plaque and its relationship to the tangle previously noted. But he also argued, perhaps a bit defensively, that new techniques did not make clinical observation and traditional histochemical methods obsolete. Alzheimer, he pointed out, was both a clinical psychiatrist and a neuropathologist and used both types of observation in his first description of the disease. Light microscopy retained an advantage in being able to generate an overall picture of the location, distribution, and diversification of the changes within the brain and would remain useful for quantitative studies that could complement the highly detailed qualitative observations produced by electron microscopy. Roth summed up his thoughts about bringing together researchers who employed a variety of techniques to study Alzheimer's disease by quoting Freud's assertion that psychoanalysis and brain psychiatry should not be seen as contradictory: "I have no inclination to keep the domain of the psychologist floating as it were in the air without any organic foundation. Let the biologists go as far as they can and let us go as far as we can," Freud had said. "Some day the two will meet." In the same way, dementia research using both traditional clinical observation and histopathological methods and the work of neuropathologists employing electron microscopy needed to be vigorously pursued. The symposium demonstrated, according to Roth, that "in this important group of organic disorders there are many meeting points and some of them are beginning to show promise as growing points."[27]

Although it is not clear whether Roth intended it, his use of Freud carries an obvious irony because, however much Roth valued heterodox methodologies within a framework that viewed senile dementia as an organic disorder, a psychodynamic theory of dementia was clearly unacceptable. If the published proceedings are any indication, Roth and the other symposium participants did not even consider the possibility of a "meeting point" between their organic approach and the psychodynamic approach that dominated American psychiatry a generation earlier. In the wake of the avenues for exploration opened by electron microscopy, researchers publishing on dementia did not even bother to refute the psychodynamic model, and by 1970 Rothschild's name had virtually disappeared from citations in the professional literature. Electron microscopy's impact on Alzheimer's disease research thus set a new agenda for pathological research, an agenda that left no room for considering the possible impact of psychosocial factors in the etiology of dementia.

Terry mapped out that agenda in a review article on dementia for the January 1976 issue of the *Archives of Neurology*. Although there were many directions worth exploring, it was crucially important to identify the specific protein of the tangle in order to organize future research. If it was shown to be a derivative of normal protein, "the search for a modifying mechanism can begin rationally. . . . If on the other hand, the protein is new to the organism, then new genetic material must be sought." The tone of the article was cautiously optimistic, listing the many areas in which progress was being made but acknowledging that the rich evidence being amassed on various issues could not be integrated into an explanation of the genesis of either the pathological structures themselves or the disease as a whole. "There are, nevertheless, ample organic explanations—for what they are worth—for this most common dementia."[28] In the context of his review, it is hard to understand why Terry used the modifier "organic." His citations were almost entirely restricted to publications that followed the 1963 studies and did not, needless to say, include any reference to the psychodynamic model of the previous generation. There was simply no evidence for even the possibility of an explanation that was *not* organic. That gratuitous modifier, positioning organic explanations against some unspoken and scarcely imaginable alternative explanation, was almost all that remained of the psychodynamic model of senile dementia in the new world of Alzheimer's disease research that took shape in the 1970s and 1980s.[29]

Although Terry's review article can appropriately be read as a triumphal account of the return to dominance of a strictly biological approach to dementia, focusing on recent progress and especially evidence generated by ultrastuctural studies, it also revealed the difficulty faced by neuropathologists who endeavored to construct an organic account of Alzheimer-type dementia. If the abundant evidence produced by the electron microscope helped to reestablish the dominance of an organic approach to dementia, it also produced an embarrassment of riches. Neuropathologists could choose any number of avenues to explore, but without a comprehensive etiological theory or at least a model of the proximal factors leading to the development of pathology, they would find it difficult to organize it in a rational manner. Until a solid theory of pathogenesis or etiology was established, researchers might wander down an endless number of blind, albeit interesting, alleys. As Terry clearly indicated, lack of such a model made certain problems central—such as determining whether the neurofibrillary tangle was an alteration of normal neuronal proteins or the introduction of new material. But the importance of this question hinged on whether the tangle was, as Terry argued, the most important pathological feature of Alzheimer's disease. Other researchers made a strong case for the centrality of the plaque.[30] However rich the evidence produced by new techniques such as the electron microscope, research could not proceed on a rational basis without a substantive causal model. To play off of Roth's use of Freud again, neuropathologists had evidently gone as far as they could without a major theoretical innovation.

"SEEK SIMPLICITY AND DISTRUST IT": THE CHOLINERGIC HYPOTHESIS OF ALZHEIMER'S DISEASE

About the time Terry published his review article, Alzheimer's disease research was being dramatically reshaped by the entrance of biochemistry. A number of etiological theories were by then being researched, including the possibility of a slow virus such as the viruses that caused scrapie in sheep or kuru in humans, the toxic effects of aluminum in the brain, genetics, and the deficit of the neurotransmitter acetylcholine. Although molecular biology would come to dominate Alzheimer's research in the 1990s, the most important area of research through the 1980s, and the one to truly create the boom in Alzheimer's disease research, was this last area. The observation that the brains of patients who had died with Alzheimer-type dementia had a pronounced deficit of acetylcholine (a neurotransmitter that had been shown to play an important role in memory and learning) allowed researchers to construct a persuasive model of the proximal cause of dementia—the cholinergic hypothesis. In its most precise form, the cholinergic hypothesis argued that a single aspect of the global cognitive deterioration of Alzheimer's disease, the impairment of recent memory, was caused by a pronounced depletion of acetylcholine in the brain, which in turn was caused by the death of a specific group of neurons. Articulated this way (as it typically was in the scientific literature), the model said nothing about either the other cognitive deficits of Alzheimer's disease—apraxia, aphasia, and agnosia—or disturbances of personality and affect. Nor did it say anything about the etiology of the disease, that is, what *caused* the deterioration of this specific population of neurons. But more expansive versions appeared as well, in which memory impairment seemed a synecdoche for the global cognitive deterioration of dementia and the complex totality of Alzheimer's disease reducible to the depletion of acetylcholine.[31] Articulated this way, the cholinergic hypothesis could clearly be seen as an attempt to reduce the complex totality of Alzheimer's disease to something that could be managed by scientific research.

From the outset, the clinical relevance of the cholinergic hypothesis was obvious. If cognitive deterioration in Alzheimer-type dementia could be linked to the pathological deficit of a specific neurotransmitter, one could rationally pursue drug treatments designed either to restore the neurotransmitter to normal levels or to prevent its depletion. Such an approach had worked in the case of Parkinson's disease, in which a deficit of the neurotransmitter dopamine had been found and treated. Thus, the fate of the cholinergic hypothesis as a major research program hinged on the degree to which it could be translated into effective treatments. By the 1990s, the failure of various cholinergic treatment strategies to make a significant clinical impact dampened the enthusiasm, and the cholinergic hypothesis was overshadowed by the discovery of genes for Alzheimer's disease. But for a decade after the initial reports, the cholinergic hypothesis was the dominant line of research, attracting many new researchers and much excitement to the field of Alzheimer's disease.

Interest in the role of the cholinergic system (the "network" of neurons dispersed widely throughout the brain which process the neurotransmitter acetylcholine) in dementia was sparked from a variety of sources. By the 1960s, it was well established that drugs blocking cholinergic activity in the brain produced dementia-like disruptions of short-term memory and learning. The effects of one such drug, scopolamine, were well known because it had been used since the early twentieth century as a form of anesthesia in childbirth. Administered with a small dose of sedative or hypnotic drugs, scopolamine induced a "twilight sleep" in which women had little, if any, memory of the delivery. Experiments in animal psychopharmacology on the effects of cholinergic-blocking and cholinergic-enhancing drugs also produced evidence that the cholinergic system was an important mediator of learning and memory. Studies of the effect of brain lesions on memory and learning implicated areas of the brain, especially the hippocampus, which were rich in cholinergic neurons. Finally, the cholinergic deficit in Alzheimer's disease brain tissue had actually been reported by Alfred Pope and colleagues in 1965.[32] But this early study was not regarded as decisive evidence of a cholinergic deficit because of problems with its methodology and, more important, the lack of a conceptual frame for appreciating its significance.[33]

An important link between the cholinergic system and cognitive disturbances in aging was made by in 1974 by David Drachman and Janet Leavitt of the Department of Neurology of Northwestern University Medical School in a paper reporting the similarity of scopolamine-induced (temporary) cognitive decline in unimpaired young subjects to the cognitive decline associated with normal aging.[34] Drachman and Leavitt wanted to establish more concretely the role of the cholinergic system in memory and learning, and they used the effects of scopolamine on the memory test scores of normal young subjects as a measure. As expected, the test scores of those who had been given scopolamine were significantly lower than those of age-matched controls. Interestingly, they also compared the scores of the young subjects who had been given the drug with those of a group of elderly subjects and found that the pattern of memory and cognitive deficits was strikingly similar. Both the drugged and older subjects had significantly lower scores than the controls on tests for recent memory and learning while immediate memory and verbal IQ were intact, and both groups also exhibited a similar pattern of changes on electroencephalographs. Moreover, the Newcastle study of the brains of elderly without dementia found that, although plaques and tangles were dispersed throughout the brain, there were more of them in the acetylcholine-rich hippocampal sections of the brain. In light of this evidence, Drachman and Leavitt concluded that there was a definite—though as yet unexplained—association between deficiency in the cholinergic system and mental deterioration associated with aging.

Before describing the research that confirmed the cholinergic deficit in dementia that Drachman and Leavitt's study suggested, it is worth pointing out that detection of acetylcholine (or any neurotransmitter) in the human brain is extremely difficult. As mentioned previously, the cholinergic deficit had first been directly observed by Pope and colleagues in 1965, but methodological problems mitigated the significance of these findings. Acetylcholine in the brain cannot be directly detected in living or postmortem tissue; its presence must be inferred through the measurement of the activities of enzymes to which it is functionally related. Pope employed the enzyme acetylcholinesterase, which metabolizes excess acetylcholine after synaptic transmission, as a marker for acetylcholine. But acetylcholinesterase is present in both pre- and postsynaptic membranes and thus is not a specific marker for acetylcholine-producing neurons.[35] This was but one problem of many in performing neurotransmitter assays. Biopsy tissue from living patients, often thought of as the best material on which to conduct bio-assays, was unsuitable for the study of neurotransmitters because only a small sample of brain could be obtained from a single region, making it impossible to assess the typicality of the findings for the entire brain.[36] Using carefully selected postmortem tissue solved this problem but introduced many other complicating factors that had to be reckoned with, such as the time of death (enzyme activity fluctuates with bodily circadian rhythms), delays in obtaining tissue samples (enzyme activity decreases rapidly after death), and the temperature of the body at death and time between death and refrigeration (enzyme activity fluctuates with temperature). Variations in the condition of the patient before death, such as prolonged hospitalization or immobility (the typical course for patients with de-mentia), the effects of drug treatments, or terminal illnesses, could all influence the results of neurotransmitter assays as well.[37] Moreover, the assumption at the heart of the neurotransmitter assay—that the existence of a substance can be inferred from the measurement of the *activity* of another substance presumed to have a functional relationship to it—raises interesting epistemological issues.[38] But within the scope of this book, it must suffice to acknowledge that the results I will describe in unproblematic terms are likely to conceal interesting contingen-cies that would be well worth exploring elsewhere on their own terms.

In any event, a cholinergic deficit in Alzheimer's disease appeared to be con-firmed by 1977 when three neurochemists, Peter Davies, David Bowen, and Elaine Perry, working simultaneously in three different laboratories in the United King-dom, reported finding a significant cholinergic deficit in postmortem Alzheimer's disease brain tissue.[39] The three employed virtually the same methods—the use of choline acetyltransferase (CAT), the enzyme necessary for synthesis of acetyl-choline within cholinergic neurons, as a more reliable marker for acetylcholine, combined with steps to prevent ante- or postmortem artifacts. Elaine Perry, who had joined Bernard Tomlinson's pathology department at Newcastle in order to add a neurochemical component to the meticulous clinical-pathological correla-tions that had been performed,[40] clearly articulated the potential clinical relevance of the cholinergic deficit. In its initial publication in 1977, Perry's group pointed out that the cholinergic deficit in Alzheimer's disease was analogous to the dopa-mine deficit found in Parkinson's. In the case of Parkinson's, the dopamine deficit was successfully countered by treatments with the precursor enzyme L-dopa. After reporting their observation of a significant cholinergic deficit as compared with controls, Perry and colleagues concluded that the indirect evidence from animal studies of the effect of selective cholinergic drugs combined with the direct evidence from biochemical studies of Alzheimer's disease brain tissue "should encourage further research into the cholinergic system in senile dementia and may provide a basis for therapeutic regimens which selectively act on the cholinergic system."[41] A year later, duplicating the methods used in earlier studies to count plaques and quantify the degree of dementia, the group published a study of fifty-one cases (eighteen with dementia of the Alzheimer type, thirteen normal controls, and twenty controls with other mental disorders) showing a significant correlation between the reduction in CAT activity, the extent of pathological damage, and the degree of mental impairment. These findings suggested to the authors "a 'cholinergic hypothesis' of SDAT [senile dementia of the Alzheimer type] in which functional abnormalities are postulated to be associated with defective cholinergic neurotransmission."[42]

The cholinergic hypothesis was significantly strengthened in the early 1980s when an anatomical correlate for the cholinergic deficit was established. Peter Whitehouse and colleagues at the Johns Hopkins University School of Medicine linked the selective destruction of a specific population of subcortical neurons in the basal forebrain—the nucleus basalis of Meynert (nbM)—to the decrease in cholinergic activity throughout the cortex. Their attention was drawn to the basal forebrain by anatomical studies indicating that neurons in this small subcortical region of the brain projected throughout the cortex and that these neurons were cholinergic; in addition, animal studies showed that experimentally produced lesions to corresponding areas in the brains of rats produced a reduction in cholinergic activity which parallels that of Alzheimer's disease.[43] In a study published in 1981, the Hopkins group did a postmortem comparison of the nbM in an Alzheimer's disease case and a control. The Alzheimer's disease case showed a 90 percent loss of neurons in the nbM compared with the control, and some of the cells that did remain contained neurofibrillary tangles and other forms of degeneration.[44] A year later, the group published a study of five cases that confirmed the selective destruction of neurons. The average decrease in the neuron count in the nbM was 79 percent, and the decrease in nbM cell density 73 percent.[45]

A major part of the excitement generated by the cholinergic hypothesis and especially the discovery of a specific lesion in the nbM was that this new information seemed to reduce the daunting totality of Alzheimer's disease to something more manageable. As Whitehouse and colleagues concluded in their 1982 article in *Science*, the nbM lesion in Alzheimer's disease was "the first example of loss of a transmitter-specific cell population in a major disorder of higher cortical function and, as such, represents a significant step in the understanding of the pathophysiology of this neurological disorder."[46] In a news article in *Science* reporting on the nbM research, Donald Price of the Hopkins group expressed the sentiment less formally. In explaining why scientists had long overlooked the nbM, Price observed that the disease appeared to be too complex and to affect too much of the brain to warrant consideration of a specific population of neurons: "We would look at the brain and see all these changes and say. 'Oh my God, is there any selectivity. . . . [Now] for perhaps the first time we can focus on a cell population that we know is affected by Alzheimer's disease."[47]

The cholinergic hypothesis, of course, had obvious limits. From the outset, there was evidence that other neurotransmitter systems were involved in Alzheimer's disease, though these were not as consistently shown or as dramatic as the cholinergic deficit.[48] As Price put it, "the cholinergic system may be the tip of an iceberg."[49] Moreover, memory was but one of many cognitive and mental disturbances, though it was regarded as perhaps central because it was usually the first clinical indication of a developing dementia. Finally, although the cholinergic deficit perhaps provided an important clue to the etiology of the disease, it was never developed into a comprehensive etiological theory. Why the nbM neurons were selectively destroyed in Alzheimer's disease (AD) remained a mystery.

Although these limitations were duly noted by those who developed the cholinergic hypothesis, they were sometimes elided in the scientific literature of the early 1980s to suggest that the elements of the cholinergic hypothesis were the central features of the disease. Thus, though other neurotransmitters were clearly, if less dramatically, involved, such that the cholinergic deficit might be "the tip of an iceberg," the title of an article by Joseph Coyle and other members of the Hopkins group suggested something more conclusive: "Alzheimer's Disease: A Disorder of Cortical Cholinergic Innervation." The authors acknowledged that evidence supporting the cholinergic hypothesis "does not mean that the cholinergic projections from neurons in the dbB [diagonal band of Broca] and nbM are the only systems affected in these types of dementia," but two paragraphs later, when comparing Alzheimer's disease with Parkinson's disease, they nonetheless imputed causal weight to these lesions by noting that "both disorders appear to *result primarily* from the loss of a relatively small population of transmitter-specific neurons with cell bodies located in the base of the brain."[50]

Similarly, though the cholinergic deficit had been directly linked to only one clinical symptom—the impairment of memory—researchers often suggested a connection to the global cognitive deterioration of dementia. This connection was first suggested by Elaine Perry and the Newcastle group when they found a significant correlation between the depletion of CAT and the measures of dementia developed in their earlier correlation studies.[51] The study did not make a direct connection, however, because it did not isolate the effects of the cholinergic deficit from other pathological processes that might contribute to dementia in these patients. When, through studies of the effects of scopolamine or the experimental destruction of cholinergic neurons in animals, the cholinergic deficit was isolated from whatever else might be going on in the Alzheimer's disease brain, the observed deficits were limited to memory functions.[52] But the distinction between memory loss and the global cognitive deterioration of dementia was consistently blurred in the scientific literature on the cholinergic hypothesis. Sometimes researchers simply failed to problematize the issue (as was the case with the study by Perry and colleagues), suggesting, for instance, that the loss of cholinergic nbM neurons had been linked to "the severity of clinically apparent dementia."[53] At other times, the elision was more subtle, acknowledging that the cholinergic deficit was directly linked only to memory impairment but assuming that memory impairment was somehow linked to global cognitive, if not mental, deterioration. Thus, a researcher might assert that the disturbance of cholinergic brain function plays "an important role in the memory loss and related cognitive problems associated with old age and dementia."[54] A review article might describe research of "neurologists and neuroscientists concerned with the relationship between disorders of the central cholinergic system, and the impairment of memory and cognitive (M/C) function that occurs with aging and dementia," and continue to use the elision "M/C" even though the research described was restricted to the study of memory impairment.[55]

Such conflations were ubiquitous, suggesting the adage that if you give someone a hammer, every problem begins to look like a nail. Although earlier descriptions of Alzheimer's disease had given equal weight to other cognitive symptoms and even to disturbances of personality (Alois Alzheimer's first description of the disease in fact listed paranoid delusions as the first indication of dementia), memory impairment became the primary symptom of dementia in the 1980s. The stereotypical AD case began with impairment of memory and proceeded to higher functions. "Generally, the onset of senile dementia is heralded by impairments in recent memory," Coyle and other members of the Hopkins group asserted. "Inevitably, higher cognitive functions deteriorate and the patients lose the ability to read, write, calculate, or use language appropriately." (It is interesting to note, however, that in the case in which the Hopkins group first described the loss of neurons in the nbM, impaired judgment was noted along with memory loss as the first indication of the disease which eventually brought the patient to the clinic for a neurological exam.)[56] The elevation of memory loss as the primary symptom of dementia was even more graphic in popularizations. Richard J. Wurtman, a neuroscientist at the Massachusetts Institute of Technology, began an account of

Alzheimer's disease in *Scientific American* by noting that "every year several hundred thousand Americans begin to lose their ability to remember whether they have turned off the stove or locked the front door." Although they "continue to look well" in the early stages and show no signs of other disorders that would produce such symptoms, they continue to deteriorate, and "within 3 to 10 years they will be severely demented: in effect, deprived of reason. They will be unable to speak or think or take care of themselves, and in time they will die of some complication that afflicts bedridden patients."[57] In such descriptions, memory impairment is not only the first symptom to appear but is linked to other cognitive deficits in a unified and inexorable course of mental deterioration. The anxiety biomedical discourse generated by freighting a common and usually benign accompaniment of old age—mild forgetfulness—with the totality of dementia is discussed in chapter 5.[58] Within biomedical discourse, the use of memory as a synecdoche for dementia simply brought much more of the disease under the explanatory umbrella of the cholinergic hypothesis.

This is not to suggest that the architects of the cholinergic hypothesis were deliberately or deceptively trying to inflate their claims. The articles from which the elisions were taken also contained more measured passages in which the limitations of the cholinergic hypothesis were made clear. The slip to more expansive formulation indicates how hard they strove to reduce the complexity of Alzheimer's disease to a limited problem for which one could create a satisfying and therapeutically meaningful explanation. Whitehouse and colleagues defended the interpretive boldness of the cholinergic hypothesis in a 1985 publication ambitiously titled "Dementia: Bridging the Brain-Behavior Barrier." They noted that the neurosciences had developed an impressive array of techniques to describe the clinical appearance of dementia and the postmortem appearance of the brain. But "without specific models to relate brain and behavior dysfunction, there is danger of creating data without direction," they argued. Models had dangers of their own, of course, and they should be considered no more than "simplified attempts to understand selected aspects of the complex pathophysiology of the condition." The authors evidently regarded this need both to seek and to criticize simplicity as the alpha and omega of dementia research: the epigraph for the article was a quotation by the British philosopher and mathematician Alfred North Whitehead, "seek simplicity and distrust it"; the final sentence was Einstein's dictum to "simplify as far as possible but no further."[59]

Whether or not the cholinergic hypothesis oversimplified Alzheimer's disease, its continued salience as a pathognomic model hinged on its therapeutic relevance. As Whitehouse put it, "what types and degrees of cognitive impairment are due to cholinergic dysfunction remain empirical questions. In many ways, the ultimate answer to these questions will depend on whether a drug that acts through the cholinergic system is efficacious enough to be approved [by the Food and Drug Administration] for use in AD or related disorders that demonstrate loss of cholinergic markers"[60] Three strategies for developing drug therapies emerged from the cholinergic hypothesis. The first and simplest strategy was precursor loading, in which choline and lecithin (substances normally obtained through diet which are necessary for the synthesis of acetylcholine) were administered to patients with dementia in pure form in the hope of boosting the amount of acetylcholine produced by the dying cholinergic neurons. But this strategy did not yield significant improvements in cognitive functioning and was abandoned by the early 1980s.[61] Another strategy was to develop drugs that would stimulate the cortical postsynaptic cholinergic receptors directly, making up for the loss of cholinergic neurons in the basal forebrain. This strategy continues to be pursued but has not resulted in a licensed drug. The third strategy was to develop drugs that inhibit the action of acetylcholinesterase, the enzyme that metabolizes excess acetylcholine after synaptic transmission. Cholinesterase inhibitors have been the most fruitful approach. After years of wrangling and controversy over flawed initial studies, the Food and Drug Administration approved tacrine in 1993—the first drug licensed to treat the cognitive symptoms of Alzheimer's disease.[62] Although statistically significant and reproducible, the effects of tacrine were moderate at best and limited to the early stages of the disease. As acknowledged in advertisements for Cognex, Warner-Lambert's trade name for the drug, its efficacy would diminish as the disease progressed. If tacrine (and any of the other three cholinesterase inhibitors that have subsequently been licensed for treatment in Alzheimer's) acts by slowing the degradation of acetylcholine produced by intact neurons, its effects would lessen as more and more cholinergic neurons were lost.

Given the modest degree of success that cholinergic therapies have shown, it was not surprising that in the 1990s the other aspects of Alzheimer's disease research moved into the spotlight. Identification of the specific proteins composing the senile plaques and neurofibrillary tangles generated much attention, and the identification in the 1990s of four genes mediating the development of Alzheimer's disease made genetics the cutting-edge science in Alzheimer's disease research. But the cholinergic hypothesis nonetheless had continued relevance because, for the foreseeable future, it provided the only rationale for developing treatments, albeit of modest efficacy. And the ability to find treatments, however modest their effect, is an indication that a particular problem is a fit problem for the attention of modern biomedical science. In the context of the 1970s and 1980s, the cholinergic hypothesis was the key to making Alzheimer's disease research a viable enterprise in the era of "Big Science." A 1981 news article in *Science* describing progress in Alzheimer's disease research began with an anecdote related by David Drachman concerning his reaction when approached fifteen years earlier by a funder offering to support research on Alzheimer's disease. "Drachman declined, saying 'Once we find a cure for baldness, maybe we can work on what goes on inside the head.' Baldness, of course, still has no cure, but in the past few years there have been signs that the senile dementias are not so unapproachable after all."[63] Drachman's anecdote suggests that making Alzheimer's disease research a viable scientific enterprise required more than laboratory and clinical research; it also entailed creating a new discursive framework that brought it out of the realm of pseudoscience in pursuit of the trivial—a cure for baldness—and into the realm of the ongoing fight against a dread disease.

"WE CERTAINLY AREN'T GOING TO CURE AGING": NOSOLOGY AND THE DISCURSIVE FRAMEWORK OF ALZHEIMER'S DISEASE

The discursive framework necessary to make Alzheimer's disease a viable project for modern biomedical science centered on the issue of whether Alzheimer's was a distinct disease entity or an intensification of a normal process of growing old. As discussed in earlier chapters, the issue of whether senile dementia was a distinct disease entity or an exaggeration of normal aging has bedeviled researchers since Alois Alzheimer's first description of the disease in 1907. Nor was the evidence generated by the new technologies discussed in this chapter sufficient to resolve the matter. On the one hand, no pathological, biochemical, or clinical markers existed (or yet exist) which qualitatively differentiate Alzheimer's disease from normal aging. If Alzheimer's disease were a distinct disease entity, one would expect a number of these markers to be distributed in a bimodal pattern throughout the aging population, differentiating the diseased from the normal. But every potential marker for Alzheimer's disease more closely follows a linear distribution ascending with age, suggesting that Alzheimer's disease is the end point of a continuum. One could suppose, as proponents of a disease model do, that real markers have simply not been identified yet or cannot yet be measured with sufficient accuracy to reveal the expected distribution, but such suppositions are far from persuasive evidence. There are also problems in viewing Alzheimer's as but an extreme point on the continuum of aging. From the continuum model, it would follow that everyone who lived long enough would develop Alzheimer's, and it ought to be possible to establish an age limit beyond which all survivors have dementia. While the prevalence of Alzheimer's clearly climbs with age, there are many well-established cases of people living to be very old—well past a hundred years—with little cognitive impairment. If the endpoint of such a continuum is well beyond the human lifespan, conceptualizing Alzheimer's this way becomes virtually meaningless. One could suppose that individuals age at different rates, but this would lead back toward viewing those who aged "prematurely" as having a distinct disease.[64]

Although conclusive evidence might be lacking, the issue was too important to be ignored. Many researchers seemed to feel that the issue called into question the value of their enterprise. Alzheimer's disease research had to be about more than an investigation into one of the many effects of aging; to ascribe an etiological role to aging, it seemed, was to say nothing at all. Thus researchers engaged in the various enterprises described in this chapter often struggled in their discourse to disassociate the effects of aging from "real" pathological processes, metaphorically linking Alzheimer's to other diseases, concepts, and trends associated with scientific progress. This can be seen in an extract of the published proceedings of the 1969 Ciba Foundation symposium in which two participants (both colleagues of Terry's at the Albert Einstein College of Medicine) redirected strong evidence that Alzheimer's disease was connected to the process of aging—the invariable presence of neurofibrillary tangles in the hippocampus of patients over 90 years old in the Newcastle study—by drawing an analogy to cancer:

M.L. SHELANSKI: This leads one to think that there is an ageing factor in everybody and just some controlling factor which differs. An analogous case is carcinoma of the prostate. Few people would claim that ageing caused carcinoma of the prostate, yet if a male population ages for long enough the incidence is about 97 per cent.

S.H. BARONDES: Why is that not due to ageing?

SHELANSKI: I don't know what the aetiology is. In the sort of thinking we are doing about cancer today, most of us would not be willing to accept just ageing as an aetiological entity.

BARONDES: The same is true with Alzheimer's disease. It need not be 'ageing'; it could just be an invariable consequence of having lived long enough to have been exposed to something—like radiation.[65]

If we take this exchange at face value, it is difficult to understand why it was important to assert that the unknown putative etiological agent of Alzheimer's (or of carcinoma of the prostate) was different from the equally unknown putative processes of aging. But rhetorically this assertion was crucial because studying aging itself did not seem like a legitimate scientific problem. This can be seen more clearly in an exchange from the published proceedings of another major international conference on Alzheimer's disease, the Workshop-Conference on Alzheimer's Disease–Senile Dementia and Related Disorders held in 1977:

> DR. SOKOLOFF: Are these plaque formations part of the normal aging process or part of the disease?
>
> DR. ROTH: Statistically they are normal.
>
> . . .
>
> DR. SOKOLOFF: If it is due to normal aging, we certainly aren't going to cure aging.

At this point in the exchange, Robert Terry and Martin Roth take issue with Sokoloff, suggesting that the distinction between aging and disease is irrelevant given the necessity of finding some way to intervene. But Sokoloff refuses to recognize aging as a process in which medical interventions can be meaningful:

> DR. TERRY: Childbirth is a natural process, and physicians have had an ameliorating effect on it. There is no point in arguing about this business as to whether it is a disease or normal. It is a process that diminishes the well-being of the public, and if we can have an effect on it, we should do so and not worry about whether it is normal or a disease.
>
> . . .
>
> DR. ROTH: Whether it is pathological or normal aging does not influence the desire to control the phenomenon if it is correlated with intellectual impairment and, at its extreme, with gross intellectual impairment.
>
> DR. SOKOLOFF: It may be correlated with a process about which I can do absolutely nothing.
>
> DR. ROTH: You can't assume that in advance.
>
> DR. SOKOLOFF: We don't know that it isn't.
>
> DR. ROTH: What is the purpose of deciding that in advance of investigating?
>
> DR. SOKOLOFF: That brings us around to the last question, Is it a disease or is it normal aging?
>
> DR. ROTH: Would you accept senile dementia as a disease?
>
> DR. SOKOLOFF: Yes.
>
> DR. ROTH: If you accept senile dementia as a disease, then the existing evidence, such as it is, shows that it is merely an extreme variant of what you find in the brains of perfectly well-preserved people who go to their graves with plaques in their brains, nobody being the wiser. If it is an extreme variant of that sort, there

may be hope of influencing the factors that potentiate that change, and there are many factors which do: pernicious anemia, hyperthyroidism, head injury, and perhaps serious cardiac failure. If you accept that this is so, then we are justified in investigating the conditions under which the accumulation keeps you this side of the threshold.[66]

Interestingly, Roth's final comment in the exchange abandons the claim that the interventions can be meaningfully aimed at aging. Rather, investigating the pathological process, whether it is aging or disease, is legitimated by the ability to intervene in secondary disease processes that mediate the expression of the aging/disease process that produces dementia.

For researchers steeped in the ideology of the gerontological persuasion, the idea that Alzheimer's disease was in any way connected with normal aging was anathema, smacking of the ageist refusal to take seriously the medical problems of old age. There were two aspects to this problem. The first was the failure to recognize senile dementia as a major health issue that required a significant public commitment to funding scientific research. The second aspect was the carelessness with which clinicians diagnosed irreversible senile dementia in old age, leading them to overlook many treatable and even reversible causes of dementia-like symptoms.

The failure to recognize the public health implications of dementia was addressed most thoroughly by Robert Katzman, who published a highly influential editorial in the *Archives of Neurology* in 1976, which argued that the distinction between pre-senile Alzheimer's disease and senile dementia ought to be dropped because the two conditions could not be distinguished at the clinical, histological, or ultrastructural levels. For all practical purposes they were identical. By bringing the two together, Katzman was able to draw some striking epidemiological conclusions based on community studies conducted in Britain and the Scandinavian countries: 4.1 percent of the population over age 65 suffered from severe dementia and 10.8 percent from mild dementia. Extrapolating from these figures, Katzman estimated that there were between 880,000 and 1.2 million cases of Alzheimer's disease in the United States and that 60,000 to 90,000 deaths per year could be attributed to it, making it the fourth or fifth leading cause of death in the adult population. But Katzman's claims regarding the prevalence of dementia was not as important as his argument about what kind of problem it presented. After all, the problem of the increasing prevalence of dementia with the aging of the population had been discussed in psychiatric and gerontological literature since the 1930s. But by calling the unified entity Alzheimer's disease, Katzman was able to link this increasing prevalence to a disease entity with a well-defined pathological basis, that is, a disease that was demonstrably "real" and thus worthy of a massive research effort into its cause and cure. Katzman made a clear distinction in his editorial between normal aging, the "benign senescent forgetfulness" that is common in old age but which does not lead to Alzheimer's disease, and Alzheimer's disease itself at whatever age it occurs. The "senile as well as the presenile form of Alzheimer are a single disease," he concluded, "a disease whose etiology must be determined, whose course must be aborted, and ultimately a disease to be prevented."[67]

One outcome of the Workshop-Conference on Alzheimer's Disease–Senile Dementia and Related Disorders, which was previously quoted, was that Katzman's formulation was endorsed by the consensus of a community of experts working in the field. Following publication of the editorial, Katzman, along with Robert Terry, sent a letter suggesting the Workshop-Conference on Alzheimer's Disease to Donald Tower, director of the National Institute of Neurological and Communicative Disorders and Stroke (NINCDS). Based on the excitement generated by the discovery of the cholinergic deficit, Tower agreed to the idea, brought in Katherine Bick of the NINCDS to help organize the conference, and brought in the directors of the National Institute on Aging (NIA) and the National Institute of Mental Health (NIMH) to make those institutes co-sponsors.[68] The conference brought together virtually everyone doing significant work on Alzheimer's disease at the time with the purpose of defining research goals and attracting new researchers to the field. Twenty-seven of these scientists served on three working committees to formulate a set of recommendations regarding etiology and pathophysiology, epidemiology, and nosology.

The combined report of these committees, published in the conference proceedings, recommended the term *senile dementia of the Alzheimer type* to describe the condition in patients over 65. *Alzheimer's disease* would continue to be the term to describe the condition in patients under 65. Although it preserved an arbitrary distinction based on nothing more than the age of onset, which Katzman objected to in the editorial, this terminology recognized that the two were essentially the same disorder while leaving open the possibility that subsequent research would reveal a different etiological basis for dementia appearing in a 50-year-old as compared with that appearing in an 80-year-old. As Katzman put it in the general discussion section of the conference proceedings, "We may be at the stage where we now have said, 'yes, if you spill sugar in the urine you are a diabetic, whether you are young or old.' The next stage is to say that, in fact, the pathophysiology and etiology of juvenile and late-onset diabetes is different."[69]

This formulation, with minor variations, was quickly adopted in the professional literature on the disorder. Researchers either used the recommended AD/SDAT distinction or used the term *Alzheimer's disease* to describe progressive dementia whatever the age of onset.

As described in chapter 3, changing the benighted attitudes of clinicians regarding mental deterioration in old age was one of the rallying points of aging advocates such as Robert Butler. The broad concept of "senility" or "senile dementia" lumped too much together, ascribing a broad range of mental disorders in the elderly to the irrevocable effects of the passage of time; it was a wastebasket term that needed to be replaced by a precise nosology of mental disorders in old age and careful and rigorous diagnostic practice. As described in chapter 2, psychiatrists in the 1940s and 1950s had begun to express concern that various treatable or preventable conditions, such as nutritional insufficiency, drug reactions, and even the effects of hospitalization, could produce symptoms that could be mistaken for senile dementia. But at the same time, the psychodynamic model, as Martin Roth complained, combined these factors, social and psychological factors, and biological factors into a single broad concept of dementia that made diagnostic precision seem unimportant and in any case virtually impossible. In the 1970s and 1980s, this broad concept of dementia was disaggregated and sorted into discrete categories on the basis of putative etiological cause.

One part of deconstructing the broad concept of senility involved distinguishing between Alzheimer-type dementia and cerebral arteriosclerosis. The two were still frequently thought of as related processes through the 1960s. The work of Corsellis and the Newcastle group, as well as others, demonstrated that Alzheimer-type pathology and cerebral arteriosclerosis were distinct processes and that Alzheimer-type pathology was far more common. When vascular disease did produce dementia, it did so through a series of large or small strokes that caused a series of sudden, stepwise deteriorations that could be differentiated from the gradual global deterioration of Alzheimer's disease. In 1974, the term *multi-infarct dementia* was coined, replacing terms that suggested a more generalized deterioration such as cerebral arteriosclerosis or arteriosclerotic psychosis.[70]

The other part involved distinguishing between Alzheimer's disease and the various treatable and often reversible conditions that could be mistaken for Alzheimer's. In 1980, a task force on the issue, sponsored by the NIA under Butler's leadership, published a widely cited statement of the problem, listing scores of disorders ranging from depression to fecal impaction which presented dementia-like symptoms that sometimes resulted in a misdiagnosis of Alzheimer's disease.[71] Three years later, a work group sponsored by the NINCDS and the Alzheimer's Disease and Related Disorders Association (ADRDA) published an article establishing criteria for the diagnosis of Alzheimer's disease which included recommendations for diagnostic procedures to exclude other possible causes of dementia.[72] While it is difficult to know how effective these efforts have been at reducing diagnostic error, constant emphasis on the need to make a careful differential diagnosis in popular and professional literature, including the revised versions of the American Psychiatric Association's *Diagnostic and Statistical Manual of Mental Disorders* (DSM-III and DSM-III-R), seems likely at least to have made clinicians more sensitive to the issue. Currently, the National Institute on Aging claims that specialized centers diagnose Alzheimer's disease with 90 percent accuracy.[73]

In deconstructing the broad concept of senility there was a tendency to privilege brain pathology over clinical phenomenology as providing meaningful evidence of the disease. Dementia associated with a well-defined pathology seemed more real than dementia that was associated with other factors. This is clearly indicated by the term *pseudodementia,* which was used to describe functional disorders in the elderly which produced the clinical picture of dementia. The prefix *pseudo* suggests that when dementia is not produced by a well-defined pathology, it should be regarded as merely an imitation of real (i.e., organic) dementia. Some clinicians objected to this term on just these grounds. At the 1977 Workshop-Conference on Alzheimer's Disease, Marshal Folstein and Paul McHugh argued that since depression in the elderly often produced cognitive symptoms that could not be differentiated from other forms of dementia, and since these cognitive symptoms share reversibility with those of many other dementia syndromes, the term *pseudodementia* was inappropriate. "The choice of the term pseudodementia may . . . lead us to explain away as 'false' or 'pseudo' the symptoms which in every other condition lead us to find neuropathology and must here lead similarly to a neuropathology of manic-depressive disorder, a goal of importance to clinical psychiatry and the neurosciences. We prefer to employ the term 'dementia syndrome of depression' and expect a description of the neuropathology from further studies."[74] Martin Roth objected strongly to this terminology, reiterating complaints about diagnostic sloppiness which he had first made in the 1950s (see chapter 2). Roth felt that nosological categories should reflect differences in etiology and course rather than in clinical symptomatology so that "where you have a coloring of intellectual impairment that responds to treatment you have to differentiate it from a known etiology." Adopting terminology that obscured that distinction, as Folstein and McHugh suggested, would put clinical psychiatry "back with the practice that was pursued

30 or 40 years ago, because this is precisely what happened with the aged—all paths led to dementia. There was depressive dementia, paranoid dementia, confused dementia. It is rather like the medieval belief that all disease was a scourge of God, and therefore they were all one—you didn't have to make differences in diagnosis."[75]

Other clinicians claimed that the artificiality of pseudodementia was in fact detectable and could be the basis of differential diagnosis. Charles Wells, a psychiatrist at Vanderbilt University who published widely on dementia and pseudodementia in the 1970s and 1980s, formalized diagnostic criteria for differentiating the two. There were more than twenty criteria, but most of them revolved around the degree of insight and awareness a patient had of his or her cognitive deficit. Patients with a functional disorder that presented clinically as dementia typically displayed an awareness of their cognitive impairments and made no effort to conceal them—they would, in fact, often call attention to their failures, errors, and limitations. Patients with organic brain disease usually seemed unaware of their cognitive decline or made efforts to conceal or explain it away. In this conception, pseudodementia appears to be a kind of fraud or con game that needs to be detected by careful clinical observation. "Close attention to the performance of the patients in this group [those with pseudodementia] always revealed profound inconsistencies," Wells argued. For example, he cited a woman who in conversation recounted "recent and remote events in meticulous detail and in correct temporal sequence . . . [but] when items on a psychological test . . . were read to her, she insisted that she could not 'hold on to the questions long enough to get answers.' " Another example was the man who "in conversation could provide complex and specific details about his financial problems but failed on inquiry to solve mathematical problems of minimal difficulty." Inconsistencies could be found by observing behavior alone. "For example, one patient insisted that he was incapable of learning the simplest of tasks, but he repeatedly found his way to the kitchen, easily assembled desired snacks, and skillfully found hiding places just at the moment he was scheduled to perform assigned chores."[76] Wells did not intend, of course, to suggest that elderly patients with functional mental disorders were willfully deceiving the clinician. Nonetheless, in his conception, not only was their dementia less real, but their behavior was stigmatized as deceptive and misleading, if not conniving, in comparison with the authentic suffering of patients afflicted with "real" dementia.

The ascendancy of the pathological can also be seen in the question of what evidence was needed to provide a conclusive diagnosis of dementia. From Alois Alzheimer's time through the 1950s, it was assumed that a careful clinical diagnosis could be regarded as conclusive evidence of Alzheimer's disease (or Alzheimer-type senile dementia). When, as often enough was the case, the clinical diagnosis did not correspond with pathological evidence found postmortem, it was the pathological evidence that was thrown into question, not the accuracy of the diagnosis. This faith that clinical observation could be the basis of nosology and diagnosis—faith that Martin Roth argued was badly misplaced—made the psychodynamic model of senile dementia possible by creating anomalies in the pathological picture which could best be explained by bringing in social and psychological factors to explain dementia. In the 1970s and 1980s, exactly the opposite situation prevailed. If the pathological evidence found at autopsy did not agree with the clinical diagnosis, the diagnosis was presumed to be in error. According to the NINCDS-ADRDA diagnostic criteria, a definite diagnosis of Alzheimer's disease can be made only if the clinical criteria for probable Alzheimer's disease are met *and* confirming pathological evidence is obtained either at biopsy or autopsy. Although the extensive clinical criteria for Alzheimer's disease might be fully met, the diagnosis would not stand without confirming pathological evidence. (The criteria made no recommendations regarding the opposite situation, in which pathological evidence of Alzheimer's disease conflicted with the clinical observation of normal mentation).[77] Clinical observation, while still crucial to differentiate SDAT or AD from other conditions, was not regarded as conclusive evidence of the disease in the new world of AD research of the 1970s and 1980s. Demonstrable pathology had become the standard of diagnosis as well as the evidence of the reality and importance of Alzheimer's disease.

"THIS SMACKS OF POLITICS RATHER THAN SCIENCE!"

Framing senile dementia not as an extreme form of aging but as a distinct disease entity, whose clear pathological basis made it demonstrably real, had important implications for politics and policy. To be a politically viable enterprise, Alzheimer's disease research could not appear to be an investigation into an aspect of aging, for this could too easily be dismissed in an ageist society. Research on age-associated progressive dementia had to be about fighting a dread disease, Alzheimer's, a major killer—the disease of the century. As the next chapter shows, this formulation was the central dogma of policy advocates (many of whom were also scientists engaged in AD research) who waged a tremendously successful campaign to win federal support for Alzheimer's research. I conclude this chapter by briefly considering the charge that the political utility of this formulation makes it bad science.

The political relevance of the disease model of senile dementia was not lost on either its proponents or its critics. Scientists involved in the effort to win financial support for Alzheimer's disease research did not necessarily disguise the importance of the model in their efforts. Robert Katzman made the point openly at a 1988 Ciba Foundation symposium on aging research. "I have spent a number of years trying to persuade people that Alzheimer's disease *is* a disease, and not simply what used to be called 'senility' or 'senile dementia,' " he said. "And there has been marvelous progress in research. In my view, this is because people now consider Alzheimer's as a disease." Within biomedical discourse, those who have disagreed with the disease model sometimes argued that this political relevance made it illegitimate. In response to Katzman's comments, British geriatrician J. Grimley Evans retorted that "this smacks of politics rather than science!" Evans thought that, in the absence of a solid understanding of what exactly aging was, no conclusion should be reached regarding the issue of aging versus disease. "Translating 'senile dementia' into Alzheimer's disease has undoubtedly produced public and financial support for research that would not otherwise have been forthcoming. But this is merely an example of terminology being used to solve a problem that it created in the first place. If dementia had not been for so many years categorized as 'normal ageing' the relevant research might have been started decades ago."[78]

American geriatrician James Goodwin articulated the charge that this conception of Alzheimer's disease was politically motivated in an article in the *Journal of the American Geriatrics Society,* provocatively titled "Geriatric Ideology: The Myth of the Myth of Senility." Goodwin argued that geriatricians and gerontologists viewed themselves as "in a struggle against ageism, against the forces in medicine and society at large that do not value old age and the elderly. Such a struggle promotes the growth of ideology. Ideology introduces a moral context to scientific discussions. In other words, ideas or discoveries are good or bad, not just right or wrong." The central tenet of geriatric ideology, or what I have called more broadly in the previous chapter the gerontological persuasion, was that most, if not all, of the physical and mental deterioration associated with aging is not due to aging per se but to disease or other extrinsic factors. Goodwin complained that, while geriatric ideology had the good effect of increasing awareness about the problems of old age, it stifled the scientific enterprise by making it impossible to generate and test hypotheses that conflicted with central tenets of the ideology. The ideological transformation of senility and senile dementia into Alzheimer's disease was a perfect example of this. Goodwin wondered what was wrong with the hypothesis that "what we now call Alzheimer's disease . . . is the clinical expression of an aging brain." In his view,

> in a scientific context it is either right or wrong, or somewhere in between. In an ideological context, however, this hypothesis is bad. It has adverse implications. If senility is the manifestation of an aging brain, then perhaps this will encourage lack of respect for the elderly; perhaps it will discourage expansion of federal funding for research on the problems of aging; perhaps it will encourage a sense of fatalism and an avoidance of health-promoting activities in our older patients. All of these concerns are real. However, they have absolutely no bearing on whether the hypothesis is correct or incorrect.[79]

Goodwin is clearly right that Alzheimer's disease research has been ideologically driven, but his assertion that this in itself makes it bad or at least tainted science is naive. He is not alone in this. Critics of the "biomedicalization" of Alzheimer's disease research have also explicitly or implicitly suggested that because Alzheimer's disease research is clearly connected to ideological and political agendas it is illegitimate, and, even more remarkably, they seem to believe that a research program based on a psychosocial model would be free of, or at least in its reflexivity less captive to, such ideological contamination.[80] But ideological commitments are a necessary part of all science, whether social or biomedical. Such commitments allow scientists to make rational decisions about what questions ought to be asked and what data ought to be regarded as important. Without ideological commitments science would wallow in an infinite sea of undifferentiated hypotheses and data. To put it another way, the clear a priori commitment of leading Alzheimer's disease researchers to what might be called the central dogma of the gerontological persuasion—that dementia is a product of disease processes distinct from the process of aging—does not preclude their commitment to the tenets of objective science. Indeed, it may *demand* a commitment to objectivity as one of the fundamental values of the scientific enterprise.

In sum, there is a meaningful critique of the biomedical model of Alzheimer's disease, but it must be more than the demonstration that it is ideologically driven. One must consider the moral and practical implications of the biomedical model in the concrete situations in which it has been deployed. One must ask, first of all, whether the biomedical model of dementia has generated interesting and useful scientific knowledge. On that score, as this chapter has shown, the biomedical model must be regarded as a tremendous success—though it is certainly possible to argue that other formulations of the problem might have been equally productive or that the biomedical model may have played itself out as a scientific theory. One must also ask what sort of political and policy ramifications are entailed in generating this knowledge, how it has been used and by whom, and how it intersects with other kinds of knowledge (the subject of the next chapter). Finally, one must consider the type of moral world it works to create and sustain, how it plays out in the webs of meaning in which people must make their lives (the subject of both the next and the final chapters). Only when these sorts of issues have been considered can one begin to assess the worth of a scientific theory.

CHAPTER FIVE

The Health Politics of Anguish

On 15 September 1983, the U.S. House of Representatives passed a resolution declaring November of that year National Alzheimer's Disease Month. The representatives who spoke in support of the resolution recited a litany of arguments and statistics that would have been familiar to anyone remotely familiar with the professional and popular literature on aging and dementia of the previous few years. Congress passed the resolution in the hope that "an increase in the national awareness of the problem of Alzheimer's disease may stimulate the interests and concern of the American people, which may lead, in turn, to increased research and eventually to the discovery of a cure."[1]

There was nothing extraordinary about any of this, of course. The content of the speeches clearly echoed the arguments of a coalition of professionals and caregiving family members of people with Alzheimer's disease (leaders of this effort were often both caregivers and well-placed professionals) whose organized efforts were making Alzheimer's disease a household word. Moreover, talk is cheap—and such resolutions cost Congress nothing. (The Alzheimer's disease resolution was one of a handful being pushed through that day.) Within a week, President Ronald Reagan issued a proclamation making the resolution official, and in November he held a formal signing ceremony and photo opportunity in the Oval Office for leaders of the campaign to raise awareness of the disease. However prosaic, the resolution (which has been an annual political ritual ever since) was a clear indication that new meanings for old age and dementia had attained political currency.

If Reagan's signature on this resolution marked the emergence in 1983 of

Alzheimer's disease as a popular national cause, his handwritten letter in 1994 telling the nation that he himself had been diagnosed with Alzheimer's marked a high point in public awareness that would be matched only when he died a decade later. Although Reagan was 83 years old at the time, his trademark physical vigor and sunny disposition remained intact, heightening the tragedy of the diagnosis. His farewell letter was both characteristically plainspoken and elegiac: "I now begin the journey that will lead me into the sunset of my life. I know that for America there will always be a bright dawn ahead."[2] Ronald and Nancy Reagan won widespread praise for courageously and generously sacrificing their privacy in order to advance awareness and support for increased funding of biomedical research into the disease. They and other members of the Reagan family ultimately did far more than lend the Reagan name to the fight against the disease. In 1995, the Reagans launched the Ronald and Nancy Reagan Research Institute of the Alzheimer's Association to raise and distribute private funds for the research. In addition, Nancy Reagan became an honorary member of the association's board and indefatigable advocate for research, and Ronald Reagan's daughter Maureen became an active member of the board.[3]

Ronald and Nancy Reagan's decision to share his diagnosis with the public did not occur in a vacuum, of course, but fit into a strategy established by Alzheimer's advocates to put a human face on the disease—in Reagan's case, one of the best-known and widely admired faces in the world. If the reconceptualization of senility as Alzheimer's disease generated excitement among researchers, when that reconceptualization was adopted by a well-organized, articulate coalition of family members and caregivers of people with Alzheimer's, who spoke eloquently of its ravages in the mass media and congressional hearings, politicians wanted to show their support as well. By the mid-1980s, national politicians were eager to be seen as part of a crusade to eradicate this seemingly new dread disease. And, to a remarkable degree in an era in which budget deficits dominated federal policy making, they were willing to put up the money. Congressional funding for biomedical research on Alzheimer's disease increased dramatically in the three decades following the scientific developments described in the previous chapter. In 1976, federal funding for Alzheimer's research was less than $1 million. It had risen to more than $11 million by 1983, when Reagan signed the resolution for Alzheimer's Disease Month; to more than $300 million when he announced his diagnosis in 1994; and to about $700 million in 2005, the year after his death.[4]

The strategy of Alzheimer's disease advocates was to juxtapose the positive appeal of the gerontological persuasion, whose central tenet was that old age is normally (or ought to be) as healthy and productive as any other stage of life, with

harrowing descriptions of the ravages of Alzheimer's disease—and to extol increased funding for biomedical research as the solution. This strategy was shaped
by broader political currents of the 1980s and 1990s, and a complete understanding of the development of policy requires a thorough exploration of that context—
a project far beyond the scope of this chapter. What follows in this chapter is not a
history of the development of Alzheimer's disease policy but a face-value analysis
of the discourse of those who worked to enlist the nation in a fight against
Alzheimer's. This analysis reveals an internal tension in the case they made for
increased funding for the disease. Alzheimer's advocates were able to persuasively frame the search for a cure as a moral imperative, but they were less able
to articulate a demand that was equally important to many of those concerned
with the disease—that government provide financial relief for overburdened caregivers. The discourse had an additional unintended consequence—deepening the
stigma associated with dementia and aging.

"THE NAME OF THE GAME IS 'ALZHEIMER'": THE ALZHEIMER'S ASSOCIATION AND THE DISEASE-SPECIFIC LOBBYING STRATEGY

To understand the development of Alzheimer's disease policy in the 1980s
and 1990s, one must begin with what Patrick Fox, in an influential article, called
the "Alzheimer's disease movement." In some ways, the term *movement* is misleading, for it is difficult (at least based on the limited studies that have been done
to date) to distinguish the history of an "Alzheimer's disease movement" from the
history of the Alzheimer's Association, one of the most visible and rapidly growing voluntary health associations in the United States and the only organization to
represent people with the disease at the national level.[5] Alzheimer's Association
leaders like to describe the organization as growing out of a "grassroots movement" of families and caregivers affected by Alzheimer's disease, who joined with
interested, sympathetic researchers, service providers, and policymakers to generate public awareness and support. In some ways, this characterization is true.
There has always been a grassroots element to the association. It has relied on the
support of a large body of committed individuals—family members seeking help
in dealing with the burdens of the disease, who are often dissatisfied with the lack
of public understanding or support. It evolved with a relatively loose, decentral-
ized structure that gave local chapters a large measure of autonomy in developing
their own programs. But in other important ways, the association has more
closely resembled an interest group much like other disease-oriented voluntary

health associations, such as the American Cancer Society, which had emerged early in the twentieth century. From its inception, its national leadership has been dominated by high-profile figures from business, medicine, nonprofit organizations, and government who are able to work comfortably with other elites in the world of biomedical science and government. The association's relationship to political and bureaucratic power has been characterized by comity rather than confrontation, relying on the high degree of legitimacy and authority of its leaders and allies to influence the shape of policy from within the system.

Whether its activities are more accurately characterized as part of a social movement, interest group politics, or a conflict or combination of the two, the Alzheimer's Association clearly played a key role in shaping Alzheimer's disease policy in the United States. Fox argues persuasively that the association's mixed achievements in policy advocacy (successful efforts to win federal support for research; failed efforts to win federal support for caregivers) were a reflection of a conflict within the association between the interests of elite researchers and family members. The interests of biomedical scientists, following the career imperatives and reward structure of the academy, were to increase federal funding for research. Family members wanted to increase funding for research as well, but they also had a strong interest in obtaining financial relief and services to support the burdens of caring for relatives with dementia. From the outset, Fox suggests, the high status of biomedical scientists within the association put them in a position to make sure their interests were the top priority. The strategy of disease-specific lobbying the association pursued was tailored to win funds for research but counterproductive in the struggle to win support for caregivers.[6] While I find Fox's invocation of "interests" unnecessarily dismissive of more altruistic motives for researchers, the conflict he describes is real, and it is a crucial element in understanding the way the federal government has responded to Alzheimer's disease.

The idea for a national voluntary organization related to Alzheimer's disease seems to have begun with biomedical researchers and government officials. According to Fox, Robert Katzman tried to establish a lay organization in New York City as early as 1974, finally succeeding in forming the Alzheimer's Disease Society in 1978. Katzman's colleague at the Albert Einstein College of Medicine, Robert Terry, was also involved in forming the society—recruiting successful Chicago businessman Jerome Stone, whose wife had Alzheimer's disease, to sit on the lay board of directors.[7]

Robert Butler was another key figure in the genesis of a national voluntary organization for Alzheimer's disease. As the first director of the National Institute on Aging (NIA; established in 1974), Butler needed to define an area of research specialization if the fledgling institute was to compete effectively for funding with other institutes within the National Institutes of Health (NIH). Other officials within the NIH had resisted establishing the NIA on the grounds that it would involve a needless duplication of administrative costs because an adequate amount of aging research was being conducted within other institutes of the NIH. Moreover, the wider biomedical research community remained skeptical that significant breakthroughs in understanding the complex processes of aging were likely and worried that a new institute focusing on aging would divert money from more worthwhile projects. Following the successful strategy of other institutes (the most obvious of which was the National Cancer Institute), Butler felt that focusing on a specific disease would allow the NIA to overcome this resistance.[8]

For this strategy to work, Butler knew that there would have to be a broader public constituency than the scientific community to get the attention of Congress, and a national voluntary association for Alzheimer's analogous to those that existed for other diseases seemed the best way to mobilize and sustain such a constituency. Butler called for the formation of such an organization at the 1977 conference sponsored by the NIA and the National Institute of Neurological and Communicative Disorders and Stroke (NINCDS) and organized by Terry and Katzman.[9] In 1979, Butler and Donald Tower, director of the NINCDS, persuaded the leaders of seven regional organizations (three of them started by scientists who had attended the 1977 conference) dealing with Alzheimer's disease and other chronic brain impairments to meet in Washington to discuss the possibility of forming a national organization.

These seven groups joined, forming the Alzheimer's Disease and Related Disorders Association (ADRDA; later renamed the Alzheimer's Association) in 1979. But by 1980, conflict arose over the disease-specific focus of the organization. Three of the groups argued that the organization should provide support and advocacy for those struggling with all sorts of brain impairments. Katzman felt that the ADRDA's family support groups could be open to anyone but that the association's efforts to educate the public and obtain support for research needed to be focused on Alzheimer's to avoid diluting their effectiveness. Most of the ADRDA leadership evidently agreed. "The name of the game is 'Alzheimer,'" an ADRDA board member argued in a letter to Katzman in July 1980. "If they do not want to play let's call them out of order and request their withdrawal. You can sell Alzheimer but you can't sell alcoholism under the title of Alzheimer and it should not be attempted."[10] Two of the three dissenting groups withdrew from the organization later that year.[11]

At stake in this conflict, Fox argues, was whether research or caregiving would be given higher priority in the organization. If the primary goal of the organization was to increase the level of federal support for biomedical research, then the narrow focus on Alzheimer's disease was the perfect strategy. The disease-specific approach had a proven track record in negotiating the politics of the NIH, and research on specific diseases was much easier to sell to Congress than basic research. Butler called this "the health politics of anguish," emphasizing that "people don't die from basic research, they don't suffer from basic research. They suffer from specific diseases."[12] The association's tremendous success in lobbying for increased federal support for biomedical research on Alzheimer's disease bore his argument out. If, on the other hand, the primary goal was to increase the level of support for caregivers, the disease-specific strategy was counterproductive. Programs and services that would help people with Alzheimer's and their families would help people with any type of brain impairment and their caregivers, and some policies dearly sought by Alzheimer's disease families, such as long-term care insurance, would benefit those struggling with virtually any chronic disabling illness. If increasing support for caregivers was the primary goal of the association, the logical course would be to create a broader constituency of people affected by the many diseases which would have benefited from the same policies.[13]

It is possible that the leadership or the rank and file of the Alzheimer's Association (or both) consciously put funding for research ahead of support for caregiving, but I have found nothing in the association's public records (annual reports, newsletters, press releases, etc.) to suggest this. Moreover, the association's commitment to winning support for caregivers seems hard to question. For example, the association was clearly at the forefront of efforts in the late 1980s to get federally supported long-term care insurance on the agenda of national politics.[14] Nonetheless, implicit in the lobbying strategy the association adopted was a trade-off—albeit unacknowledged and unexamined—between support for research and support for caregiving. This is clearer when one examines closely the language Alzheimer's disease advocates used to make their case.

"THE ONLY ANALOGY WOULD BE POLIO": CURE TRUMPING CARE IN THE DISCOURSE OF ALZHEIMER'S DISEASE ADVOCATES

There was one potential problem for the disease-specific lobbying strategy: some might want to dismiss it as feeding the "disease-of-the-month" syndrome— the tendency of the federal government to allocate resources to study one particular disease after another at the expense of support for more productive programs in basic biological science.[15] To legitimate their claim for special consideration, Alzheimer's disease advocates pointed out that the disease was already in 1980 a monumental economic and social burden—the fourth or fifth leading cause of death among adults, afflicting about 1.5 million older people and costing the nation more than $10 billion a year in nursing home costs alone. And the burden would become many times heavier as the baby boom cohort aged. Noted researcher and popular author Lewis Thomas, who also served as a member of the ADRDA's board of directors, normally criticized the disease-specific approach to allocating resources. But in an essay first published in the popular science magazine *Discover*, Thomas argued that Alzheimer's disease was "not a disease-of-the-month but a disease-of-the-century," whose economic, social, and personal costs were so great that it required earmarking funds for research directed specifically at it.[16] The phrase "disease-of-the-century" and the emphasis on the devastating burden Alzheimer's disease would put on an aging society—an emphasis that critics have called "apocalyptic demography"—became central elements of the discourse of Alzheimer's disease advocates.

The association and its allies in science and government who joined it in lobbying for the cause ostensibly endorsed with equal vigor an increase in federal money to support both research and caregiving. But in describing the "disease of the century," they forged a link between the costs of caregiving and the need for research which implicitly (and sometimes explicitly) undermined the plausibility of arguments that could be made for major social policy initiatives to address the problems of caregivers. In making the case for research funding, Alzheimer's advocates emphasized the tremendous economic burden the disease placed on society for items such as nursing home care—costs that would dramatically increase if a treatment or cure for the disease was not found. In so doing, they also underscored the degree to which policy changes that would substantially benefit caregivers would be prohibitively expensive.

Alzheimer's disease advocates often legitimated their claims for increased research funding by citing the tremendous costs of caring for people with Alzheimer's disease. The ADRDA's first chair and president, Jerome Stone, frequently decried what he saw as the imbalance between the amount of money spent on research and the amount of money spent on care. In testimony before Edward Roybal's House Select Committee on Aging, Stone pointed out that the United States was spending on research only one-tenth of 1 percent of what it paid on care—$25 million versus an estimated $26 billion.[17] "I believe that we have to accelerate funds more in proportion such as we do in diseases of cancer, heart, and other neurological based diseases as epilepsy, m/s, et cetera," he argued.[18] Stone did not intend this reasoning to undercut the case for long-term care; other parts of his testimony make it clear that finding a way to relieve caregivers of some of the staggering financial burden of the disease was a crucial part of the association's agenda. In making the same argument in the ADRDA's first annual report (1984), he tried to link the two: "Our nation still spends 800 times more to care for our nearly three million Alzheimer's victims than it allocates for research. Federal and private insurance programs still pay little or none of the staggering costs of Alzheimer's patient care."[19] But in such a formulation, these two policy goals stand in an uneasy relationship. Caregiving is positioned as an unfortunate and unnecessary burden—the price we pay for our failure to commit enough resources to find a cure.

Sometimes research was privileged more explicitly, as when David Drachman, one of the researchers whose work generated interest in the cholinergic deficit of Alzheimer's disease (see chapter 4), summarized in the *ADRDA Newsletter* what he saw as the policy implications of the twenty-two presentations made at the 1981 National Mini-White House Conference on Alzheimer's Disease. (According to the article, these recommendations were passed along to the upcoming 1981 White House Conference on Aging.) Drachman pointed out that nursing home costs for Alzheimer's disease were then estimated at more than $10 billion whereas federal funding for research was only about $10 million (this was two years before Stone's testimony). "How can we spend a thousand dollars on maintenance for every dollar that we spend in an effort to prevent, treat or cure dementing disorder?" he wondered. "It seems unrealistic to me, and, I think, to all of us." Drachman thought that efforts should be given priority in the following order: prevention, cure, treatment, assistance, and care. "This is exactly the reverse of the way in which these things are being funded at this moment. I think we need to redress that and turn it around. *Prevention should be number one.* Care should be last, because by investing in prevention we will ultimately be able to reduce the number of individuals who are afflicted."[20]

Such formulations obscured exactly who was doing the "funding." While both Stone and Drachman cite the amount of money spent by the federal government for research, much of the billions spent on caring for people with Alzheimer's disease was paid by private individuals who had not yet spent enough of their own resources to qualify for Medicaid reimbursement. It is not at all clear that the money spent on research and the money spent on care were actually part of the same "pie"-that is, the total amount of money that the federal government could or would spend on all aspects of Alzheimer's disease. On the face of it, there was no obvious reason not to approach them as completely separate issues, each to be funded from different sources within the federal government to address different ends. But the disease-specific lobbying strategy made it seem natural to lump together all money spent on a particular disease in various programs and for various purposes. Whether there actually needed to be a trade-off between funding for research and funding for caregiving, Alzheimer's disease advocates tended to talk about it that way. And to the extent that they succeeded in their lobbying efforts, members of Congress talked about it that way too. When this happened, caregiving was inevitably placed on the back burner.

This process of marginalizing caregiving, whether deliberate or not, was most obvious when Alzheimer's disease advocates invoked medical triumphalism to strengthen their claim further. Faith in biomedical progress remained widespread in American culture. Perhaps the most eloquent and popular voice of medical triumphalism was Lewis Thomas, who, was an ally of the Alzheimer's disease cause. In an essay in his popular 1974 book *The Lives of a Cell*, Thomas distinguished three forms of medical technology: "nontechnology," which included caregiving measures that did not address in any way the underlying mechanisms of disease; "halfway technology," designed to compensate for the effects of the disease or postpone death; and "real technology," the genuinely decisive technology of modern medicine, which derived its power through a genuine understanding of disease mechanisms. Halfway technology often attracted more public attention than real technology because it was impressively complicated, expensive, and difficult to administer. Thomas thought the sophisticated electronic devices, specialized hospital units, and highly trained personnel devoted to managing heart disease a good example of the attention that halfway technology received. Real technology, such as antibiotics and vaccines, was too often taken for granted because, once developed, it was relatively inexpensive and unproblematic to administer.[21]

For Thomas, these three kinds of technology were stages in a logical progression that characterized the history of medicine and which would continue to characterize it so long as society was sensible enough to commit the necessary resources to basic biomedical research, which alone could produce the knowledge on which real technology rested. Such commitment required an appreciation of real technology even though it was often overshadowed by the visibility, expense, and complexity of halfway technology; also required was the realization that real technology would be produced only by progress in basic science. "If I were a policy-maker, interested in saving money for health care over the long haul, I would regard it as an act of high prudence to give high priority to a lot more basic research in biologic science," Thomas concluded. "This is the only way to get the full mileage that biology owes to the science of medicine, even though it seems, as used to be said in the days when the phrase still had some meaning, like asking for the moon."[22]

Medical triumphalism was a constant feature of discourse on Alzheimer's disease policy. I have not found a single instance in congressional hearings during the 1980s and early 1990s in which a witness or member of Congress questioned the assumption that biomedical science would eventually find a solution to the problem of Alzheimer's disease if it was given enough resources to do so. Indeed, many people presumed such a breakthrough was imminent. When scientists or the ADRDA leadership tried to suggest that breakthrough treatments might *not* be just around the corner, they were sometimes rebuked. For example, an article in the *ADRDA Newsletter* cautioning against unrealistic expectations aroused by the announcement of medical "breakthroughs" (in this case the first published study of tacrine) drew an angry letter to the editor in the next issue. "I was irritated by the comment in another article, 'In our hearts we know that it is not that simple (a cure for the disease) . . . and that Alzheimer's is a devastating, complex disease, whose mysteries will take time and talent to solve,' " the reader wrote. "Why would one proclaim the difficulties and time that 'breakthroughs' would take? Not all research takes years for solutions or insights. We do not need to qualify the difficulty but rather encourage the solutions with determination. . . . I do know that rampant optimism is dangerous, but 'as a man thinks, so he becomes.' "[23] Although Robert Katzman and Jerome Stone could hardly be accused of pessimism regarding the prospects for a medical solution to Alzheimer's disease, they were chided by Congressman Bill Richardson when, in testimony at 1980 hearings on Alzheimer's disease, they seemed to hedge their answers to the question of when meaningful medical progress would be made if Congress doubled funding for research. "I am not trying to appear impatient but I do want to stress the need to find a cure for this illness as soon as possible," Richardson argued. "I sense that if a national strategy was mapped out with a timetable for periodic reviews of research discoveries and funding needs for Alzheimer's disease by Members of Congress and the administration—that by working together in concert with the medical community, progress would be accelerated in the search for a medical breakthrough for this illness."[24]

The most effective use of medical triumphalism was the polio analogy, developed by Robert Butler in the first congressional hearings on Alzheimer's disease in 1980. In his testimony before the Senate Special Committee on Aging,

Butler adapted Thomas's argument about basic biomedical research to Alzheimer's disease research. The nursing home, according to Butler, was at best a "halfway technology, very much like the iron lung was before the polio vaccine." (Here Butler's commitment to an interventionist, medicalized gerontology was apparent, for even the "teaching" nursing homes he envisioned—institutions attached to medical schools for purposes of research and training in the same way that hospitals were—would more logically fall into Thomas's category of "non-technology.") Butler looked forward to the day "when we have a full technology that will allow for the prevention of Alzheimer's disease." He acknowledged that support for caregivers had to be increased but nonetheless continued to position this as subordinate to research. "While we must certainly meet immediate service needs, we must at the same time remind ourselves that today's service is yesterday's research, and that today's research is tomorrow's service. Perhaps no better example is that of the polio vaccine. It is hard for me to imagine that it was just 20 years ago, in 1960, that we had our last polio epidemic." Butler concluded the point by trying to restore a balance between research and care: "Too, we must be always conscious of the need to have both service and research occur concomitantly." But this seems flat and unconvincing next to his eloquent evocation of biomedicine's power to banish disease.[25]

Those who wanted to press the case for increasing support for caregivers at most questioned *when* these medical breakthroughs would occur and argued that more had to be done for caregivers in the meantime. This approach was taken by Janet Sainer, commissioner of New York City's Department of Aging, who testified eloquently on the need for reform of Medicare policy before the House Select Committee on Aging in 1983 hearings entitled "Endless Night, Endless Mourning: Living with Alzheimer's." Sainer began by reiterating the call for increased research funding but devoted most of her testimony to making a strong case for the need to increase support for caregivers by making reimbursable under Medicare many of the expenses incurred in providing home care for people with dementia. "It may be quite a while before science finds answers that will lead to the cure or prevention of Alzheimer's," she concluded. "Meanwhile, by providing the vital day-to-day support needed, we can lessen the tragic impact of this most dreaded disease. Then, perhaps, the title of today's hearing being 'Endless Night, Endless Mourning: Living with Alzheimer's,' can be translated to 'Living with Alzheimer's More Effectively.' "[26]

But Sainer's eloquent testimony at this hearing was overshadowed by that of Butler, who immediately followed her. Butler's testimony did include a strong, clear endorsement of changing Medicare rules to make caring for people with Alzheimer's disease reimbursable. But his main thrust was to make the case for biomedical research. "Between 1946 and 1964, the United States developed the most wondrous and largest generation in American history, the baby boom. We cannot wait until the first baby boomer turns gray in the year 2011, to declare war on senility," he argued. "In 1935, a March of Dimes was started. In 1961, less than 30 years later, the last polio epidemic occurred, the last thump-thump of iron lungs each spring. I submit that senility could fall the way polio did, if we invest now, and I would submit that research would be the ultimate cost containment and the ultimate service. We do not have polio anymore."[27]

Butler was not the only one to use the analogy to polio. Asked by Representative Matthew J. Rinaldo from New Jersey in 1985 hearings before the House Select Committee on Aging to select which funding should be given priority—that for research or that for caregiving, ADRDA board member Lonnie Wollin (a New York City attorney) found it tough to choose. "There is an immediate need. There are people calling our office who are frightened. They need respite and care," Wollin noted. "On the other hand, if you don't fund substantial research, you will have this problem for a long time. The only analogy would be polio. Had the money gone into treatment, we would have a magnificent portable iron lung, but no cure for the disease. I think the dollars that are available have to be spread out between the immediate respite care and research."[28] Wollin was clearly trying to balance the two policy goals equally, but for members of Congress the power of the polio analogy surely tipped the balance in favor of research. The polio analogy was a rhetorical trump card, making funding for biomedical research a moral imperative in its ability to relieve suffering and the most sensible fiscal policy because it would dramatically lower the costs of caring for people with Alzheimer's disease. If medical progress truly was inevitable, an assumption that no one seemed to question, then nothing could justify the failure to provide the necessary resources for biomedical research. If polio was indeed the "only analogy," then, whether Wollin explicitly said so or not, research had to be the number one priority.

Certainly, many members of Congress saw it that way. The hearings entitled "Endless Night, Endless Mourning" were intended, according John Heinz, the chair of the Senate Special Committee on Aging, to be a forum on issues of care. But the two other senators attending the hearings, Larry Pressler and Alphonse D'Amato, were clearly persuaded from the outset that research was the best way to address those issues. Senator Heinz began the hearings by noting that while "Alzheimer's disease and related research activities have been the subject of recent congressional hearings, today's hearings will focus on the issue of care."

On a more concrete level, Heinz intended the hearings to generate support for the long-term care legislation he had introduced earlier that year.[29] But in his opening remarks, Senator Pressler, who had been very friendly to the Alzheimer's cause because his father had had the disease, noted that "we will be going to Senator D'Amato, who is on the appropriations committee [and was appearing as a guest at the Aging Committee hearings because they were held in his home state of New York], for more funding for research. . . . If we could find a cure or treatment for this disease, we could save a lot of human anguish, and save a lot of money, at a time when we have a $200-billion-budget deficit."[30] D'Amato indicated that he was more than willing to help in this, pledging to work with the committee to double the amount of federal funding devoted to Alzheimer's disease research—about $22 million that year. "I would suggest that the dollars we spend here in research . . . will pay dividends thousands of times over," D'Amato responded. "It is the best kind of investment we can make. That coming from a fiscal conservative."[31] Suffice it to say that a fiscal conservative such as D'Amato was unwilling to entertain the idea of creating a massive social program to support caregivers. Many of the other witnesses did indeed make the case that support for caregiving needed to be a priority, but these arguments had to compete with Butler's dramatic testimony, which returned the focus of the hearings to research. Heinz's legislation, in any case, went virtually unmentioned in the hearings.

It is not hard to understand why members of Congress found the polio analogy so persuasive. Finding meaningful ways to financially support those who cared for people with Alzheimer's disease was likely to be a difficult and painful process, perhaps even impossible in a time of massive federal budget deficits. Providing money for research was relatively easy and inexpensive and allowed members of Congress the consolation of believing that even if terrible problems were being ignored in the short term, research would ultimately provide the solution. In the 1985 hearings in which the ADRDA's Wollin testified, the following exchange occurred between Representative Rinaldo and Kenneth L. Davis, professor of psychiatry and pharmacology at the Mount Sinai School of Medicine and head of the Psychiatry Department at a Veterans Administration hospital:

MR. RINALDO: During the testimony of some of the other witnesses, they talked about Medicare and Medicaid reimbursement as a viable method of financing care of Alzheimer's victims. Can you give us an estimate or guesstimate of what this would cost nationally?

. . .

DR. DAVIS: When we have talked about this in our laboratory, it is frightening. We have data with projections of the number of Alzheimer's patients in the United States by the year 2005.

MR. RINALDO: What does it show?

DR. DAVIS: Something over 2.5 million patients. . . .

. . .

MR. RINALDO: Can you give us an idea of what it would cost to Medicare?

. . .

DR. DAVIS: It is extraordinary. The country cannot afford this type of expenditure. If we are not going to force you to make some awful decisions, to force you to make truly inhumane decisions, that we as physicians cannot face, we have no choice but to invest in research, though there is hardly a guarantee that it will work.

MR. RINALDO: From your testimony, it seems to me that it would be almost impossible to fund that under Medicare because it would bankrupt the system. Additional research dollars in the magnitude of $100 million, those may provide more meaningful insight in the causes, possible causes of Alzheimer's?

DR. DAVIS: Yes.

Davis and Rinaldo went on to discuss innovative ways for spending such money which would be most effective in stimulating new research on Alzheimer's disease, ignoring Davis's caveat that "there is hardly a guarantee that it will work."[32]

When it came to support for biomedical research, members of Congress were not much inclined to listen to such caveats, even when they came from the experts. The politics of biomedical research policy were shaped by budgetary constraints in the 1980s. But they were also shaped by a more aggressive, populist lobbying strategy that no longer deferred to the pronouncements of biomedical experts regarding what was the best way to spend money for research. Inspired by the ability of groups such as ACT UP to have an impact on federal policy on AIDS, patient advocacy groups, even more traditional health voluntaries such as the Alzheimer's Association, were increasingly bold about demanding that medicine find a cure and find one *now*.[33] The new climate of biomedical science politics could be seen in an exchange between Edward Roybal, chair of the House Select Committee on Aging, and James Wyngaarden, director of the NIH, during 1983 hearings on Alzheimer's disease. Wyngaarden had asserted that the $4 million increase in funding for research within the NIH for the coming year was sufficient to give a "very high priority and to support the very best work that is presented to us." Roybal chastised Wyngaarden for putting budgetary considerations ahead of the moral imperative to fight Alzheimer's disease:

ROYBAL: This committee was told in testimony . . . that what is needed is more research, better training for doctors, and the establishment of a national effort to fight
Alzheimer's disease. Now . . . you are telling me that you have enough money. Is
that because there isn't a need for additional research or because of the President's
recommendation that expenses be minimized?

WYNGAARDEN: No agency head would ever say that there is enough money. But,
within the priorities for all of the activities that we must support, we feel that
research on Alzheimer's disease and other dementias of the aged is adequately
represented.

ROYBAL: I remember making an amendment increasing the appropriation for
research of cancer that amounted to $1 billion; $17 million is nothing in comparison. Either the disease is so serious that we need more work in the field or it
is not. I am surprised that your answer seems to be that we have enough money to
take care of the problem.

WYNGAARDEN: Within the many priority issues that have been addressed by the
President's budget, we have allowed for a $4 million increase in the next year.

ROYBAL: But my question was not with regard to the President's budget. My question was with regard to the gravity of the disease and whether or not it is of such
proportion that we need to put more money into it. I can understand very well your
answer.[34]

Roybal then turned the question to Butler, who acknowledged the difficulty of
Wyngaarden's position but argued that it was "the collective judgment of many in
the scientific community and certainly the judgment of those who suffer from
this disease that we are simply not, repeat, not, devoting adequate resources to
unraveling the mystery of it."[35] This exchange indicates that the decision to fund
Alzheimer's disease research hinged as much on the moral imperative to find a
cure for a dread disease as on the rational assessment of experts as to the scientific
merit and worth of that research in the grand scheme of biomedicine. Paradoxically, the displacement of biomedical scientists as the arbiters of research policy
by more activist patient advocacy groups seems to have increased the faith such
groups have in the power of biomedicine to solve human problems and intensified the demand for biomedical interventions.

That faith in biomedical progress provided a reason for failing to create policies to address the burdens shouldered by caregivers was perhaps not surprising.
Neither was it surprising that the lobbying strategy and discourse of the Alzheimer's Association, whether its members were aware of it or not, privileged in-

creasing the funds available for biomedical resource over increasing support for caregivers. Ultimately, both members of Congress and Alzheimer's disease advocates were simply following the larger course of the river. In the 1980s, Americans had an abiding faith in biomedicine's ability to provide a painless technical fix for problems of disease and dependency. Public confidence in large-scale programs to address such problems through the redistribution of wealth or tax-supported provision of services was much more problematic. In such a political climate, it was relatively easy to support research for a dread disease; it was difficult to preserve social programs that did exist and practically impossible to create an expansive new one.

"MANY OF THESE VICTIMS WERE SAID TO BE BRILLIANT MINDS": EXEMPLARY CASES OF ALZHEIMER'S DISEASE

In addition to testimony concerning the tremendous economic and social burden of the disease, the Alzheimer's disease movement's claim to public resources was legitimated by graphic images and descriptions of the suffering the disease caused at the personal level. Giving the disease a human face provided an emotional appeal that helped raise the moral stakes in battling the disease above the arcana of federal budget policy. As Representative Rinaldo argued at a congressional hearing, "We are not doing this just to save money in nursing homes. We are doing it to save lives and to save the dignity of millions of elderly victims and their families who have fallen to this disease."[36]

Descriptions of people with Alzheimer's disease aimed to reinforce two points that were crucial to the legitimacy of the cause: first, that Alzheimer's disease was an inexorable, relentless killer that caused intense suffering and anguish for its victims and the family members who cared for them; second, that this was not simply a matter of "going senile," for the disease struck people in their prime, people who, whatever their age, were active and involved in life. These images obviously instantiated Katzman's conceptualization of senility as Alzheimer's disease—a major killer that required the same level of public awareness and commitment as other dread diseases. Katzman's conceptualization elevated Alzheimer's disease from a rare condition to a major threat in the abstract realm of medical theory; representations of its victims were necessary to make this formulation real in the arena of popular discourse, to which members of Congress were highly sensitive. Stereotypes of individuals suffering from the disease, either directly or indirectly, as in the case of caregivers, were thus a crucial part of the success of the "health politics of anguish."[37]

These representations were especially powerful because they were based on the testimony of caregivers of people with dementia, who told their stories in congressional hearings, the mass media, and publications of the Alzheimer's Association. Whereas the earlier stereotype of senility had been largely the product of professional discourse, the stereotype that emerged in the Alzheimer's disease movement was a product of the narratives of laypeople directly affected by the disease—itself an indication of the changed social position of the elderly. This is not to say that these narratives provide direct access to the "real" Alzheimer's disease experience. Caregivers did not make sense of their experience in a vacuum. As they struggled to understand what they and the people they cared for were going through, they did so in the context of the Alzheimer's disease movement and the broader effort to transform the meaning of old age in America. Jaber Gubrium's ethnographic study of Alzheimer's disease shows that caregivers often came to understand their experiences through their participation in support groups sponsored by the Alzheimer's Association and through their encounters with professionals. But giving meaning to the experience of Alzheimer's disease in these contexts was not a simple, top-down process in which caregivers absorbed the point of view of professionals and ADRDA advocates. Gubrium shows that caregivers sometimes challenged the meaning that professionals tried to ascribe to their experience.[38] Moreover, it is clear that professionals often adopted language and images used by caregivers in their representations of the disease, which in turn shaped the experience of new groups of caregivers in recognizing and describing the problems they confronted. For example, psychiatrist Carl Eisdorfer credited Warren Easterly, president of an Alzheimer's support group in Seattle, with coining the phrase the "36-Hour Day."[39] In an interview with Eisdorfer for a documentary film, Easterly spontaneously used the term to describe his experience of caring for his wife, and the filmmakers subsequently adopted the phrase as the title.[40] The term was frequently used in popular discourse on Alzheimer's disease to describe the strain of caregiving. It ultimately became the title of one of the most popular advice manuals for Alzheimer's disease caregivers, which was written by professionals.[41] Because the meaning of dementia was worked out in these broader discourses, personal accounts of the disease share similar language and images. The cumulative effect of the personal accounts and descriptions was thus not simply a collective account of "the Alzheimer's disease experience" but a new stereotype of the person with dementia as victim of dread disease. By the same token, to say that the caregivers who shared their stories helped to create a stereotype is not to say that there was anything false or inauthentic about their suffering. Nor is it to say that their stories were co-opted by leaders of the Alzheimer's disease movement to further the cause in the arenas of politics and the mass media. Rather, the emergence of this stereotype in caregiver and professional accounts of Alzheimer's embodied the changing meaning of aging and dementia in American culture.

As people who were afflicted with a "dread disease" rather than simply affected by the generalized process of aging, the exemplary Alzheimer's disease victims who appeared in the discourse of the movement were intelligent, vigorous, and active before the disease struck. Most typically, they were at the pinnacle of, or retiring from, successful middle-class careers. If they were retiring, it was not to a bleak, boring life but to active leisure and involvement with the world. In 1980 testimony before Congress, psychiatrist Eisdorfer complained of a widespread misconception concerning whom the disease affected. "We have a classic notion of what the disease is, and unfortunately, we have a stereotype. It is—and I hate to be sexist, but because there are more older women than men, it is usually sort of a little old woman who is doddering around, sitting in a geriatric chair, not knowing time, place or person," he argued. "This is not the way we see the disease. We have the disease in one engineer [about 62 years old] who still, after two and a half years, shoots golf in the eighties and wins tennis cups."[42]

In terms of epidemiology, of course, Eisdorfer had it exactly backward: Alzheimer's disease was clearly related to age, and people with the disease who were under 65 were a small minority of the total cases (this was why Alzheimer's disease was regarded as rare when the term was restricted to those who developed dementia before age 65). Whether the "typical" person with Alzheimer's disease was "doddering around" or "sitting in a geriatric chair" could be debated, but Alzheimer's disease certainly occurred most often among older people, who were increasingly vulnerable to a number of disabling conditions. But Eisdorfer's statement that "this is not the way we see the disease" is accurate in a way that he did not intend. The case Eisdorfer cites to refute what he sees as the misleading stereotype of the disease was in fact the "classic notion" within the Alzheimer's disease movement of who gets the disease. If the little old woman in the geri chair was a common stereotype of dementia that had been around for a long time, the engineer forced by diminishing cognitive skills to quit work but still able to play a good game of golf or tennis was certainly an emerging stereotype that had increasing currency in popular discourse. The stereotype of the engineer was "the way we see the disease" when we want to see it as a dread disease rather than senility. Representations of disease victims by Alzheimer's advocates legitimated the cause by showing that the disease affected people who were productive and prominent members of society.

Eisdorfer's generalization about the type of people affected was echoed in congressional testimony by Dorothy Kirsten French, a professional opera singer in Los Angeles whose husband, a prominent neurosurgeon, had developed Alzheimer's disease. Recounting the response she received after telling her story to the *Los Angeles Times*, she described hundreds of letters from Alzheimer's disease families. "Many of these victims were said to be brilliant minds," she noted. "Lawyers, doctors, successful businessmen and, like my husband, fit individuals. It is up to us to bring purpose to the loss of these great minds." Later in her testimony, when questioned by Senator Heinz concerning the burdens of caregiving, she described the hesitation she felt in putting her husband in a nursing home. "The places I have looked at I would not put my husband in," she said. "Alzheimer's disease people should not be, I believe, in a rest home where there are many people all crippled up, poor things, at the end of their life. These kinds of things, to see for an Alzheimer's person, I believe, is very disturbing."[43]

In part, this representation reflected a tendency to justify research into health problems associated with aging by showing that they did not affect older people alone. For example, although the incidence of most forms of cancer clearly increase with age, cancer was not represented as a problem of aging in the War on Cancer.[44] Although the relationship between Alzheimer's and aging was much tighter than in the case of cancer[45] and "apocalyptic demography" was a frequent trope of discourse around policy, advocates for a war on Alzheimer's disease frequently pointed out that the effects of the disease went well beyond the mostly elderly people who developed it. In the same speech supporting Alzheimer's disease research quoted at the beginning of this section, Representative Matthew Rinaldo made explicit the tendency to legitimate the disease in terms of its effect on the young: "It is incumbent upon Congress to begin an all-out attack to find a cure for this disease. For too long, it seems to me, it has been a silent killer that has been pushed aside as a product of old age. . . . Today, however, we know that that is not true. Alzheimer's is a disease and it deserves a strong and committed effort to eradicate it. . . . We are not just fighting for the elderly when we talk about fighting Alzheimer's, we are fighting for their children and relatives, people who have a higher risk toward this disease."[46]

This tendency can also be seen in the idea that family members struggling with the physical, emotional, and financial burdens of caring for a spouse or parent with dementia were the second victims of the disease. A particular concern was the "sandwich generation," adult children (mostly women) who were caught in the dilemma of having to care for both their own children and their aging parents.[47] Stuart Roth, chair of the board of directors of the national Alzheimer's

Association, brought these ideas together in an op-ed piece in response to Reagan's diagnosis, arguing that Alzheimer's disease had been ignored because it "primarily afflicts older Americans." But, he pointed out, "ignoring 'Old-Timers' Disease' also overlooks the fact that as much as 10 percent of cases are people in their forties and fifties. Moreover, many adult children are caregivers of elderly parents. There are many examples, as well, of young children caring for grandparents after school, and even forfeiting the money set aside for their college education to caregiving. Alzheimer's is a *family* disease."[48]

Representations of Alzheimer's disease in the mass media also tended to focus on the middle-aged as exemplary cases. For example, *Newsweek*'s 3 December 1984 cover story, "The Agony of Alzheimer's Disease," pointed out that it affected people in their seventies and eighties to an overwhelmingly greater extent than people in their forties and fifties. Nonetheless, the three sidebars to the main story describing exemplary cases were devoted to people with the disease in their forties and fifties—Bill, "a handsome 57-year old Harvard graduate"; Elizabeth, a 46-year-old "educated woman" who was a supervisor at the phone company; and Carl, a 55-year-old who had been a "successful attorney" until Alzheimer's struck.[49]

There was clearly more involved in representing exemplary victims of dread disease rather than a person's age. Assumptions about gender, ethnicity, and especially class shaped representations of Alzheimer's disease as they had shaped earlier representations of senility.

The image of people with Alzheimer's disease was much less overtly gendered than earlier stereotypes of senility. While Eisdorfer's opposition between the doddering old woman (Alzheimer's as senility and hence a less devastating problem) and the golf-playing engineer (Alzheimer's as dread disease and hence a greater personal and social tragedy) suggests the continuation of the tendency to see dementia as essentially a problem of men, or at least a problem that was important to the degree that it affected men, this was in fact unusual in the discourse of the Alzheimer's disease movement. Women as much as men were clearly represented as as examples of people with the disease. Actress Rita Hayworth was the most famous person with Alzheimer's disease through the mid-1980s, eclipsed only when Ronald Reagan announced his diagnosis in 1994. The authors of the popular caregiver manual *The 36-Hour Day* made an explicit point of alternating gender. "Both men and women suffer from these diseases," they wrote. "To simplify reading, we will use the masculine pronouns *he* and *his* and the feminine pronouns *she* and *hers* in alternating chapters."[50] A similar book (also very popular) coauthored by Eisdorfer was equally careful to use gender-neutral language in discussing hypothetical cases and used men and women about equally in concrete examples.[51] To be sure, there were differences in the exemplary Alzheimer's disease victim as woman and as man which followed broader stereotypes and biases of gender. When the example was a woman, the loss of attractiveness and "charm" figured as a major part of the tragedy, and the disease was shown disrupting activities that were centered around homemaking and nurturing. When a man was used as an example, the loss of intellect and that of physical vigor were more often represented as the most tragic losses. But absent was the tendency in earlier representations of senility to see these as differences that affected the essential nature of the disease, making it more a man's problem than a woman's.

Ethnicity played a role in the representations of Alzheimer's disease as well, primarily in the way it was linked to assumptions about class. People of color were virtually invisible in representations of people with Alzheimer's disease throughout the 1980s. In both the annual reports and the newsletter of the Alzheimer's Association, virtually all the images used to depict people with the disease were Caucasian.[52] There was no overt racism in these representations, as there had been in some earlier representations of senility, and by the 1990s, images in the association's publications seem to have been carefully chosen to represent Alzheimer's disease as a problem for people of every ethnicity. But when images of people of color were used, their clothing and surroundings clearly identified them as part of the mainstream middle class—as in photographic representations of people with Alzheimer's disease who were Caucasian. It was evidently not important for the legitimacy of Alzheimer's disease that exemplary victims be "white," but it was crucially important that they be respectable, productive members of the middle class. The absence of people of color from the iconography of Alzheimer's disease in the 1980s likely reflected the assumption of middle-class whites, usually unspoken, that most nonwhites were in the lower class. More careful photographic representations in the 1990s depicted people with Alzheimer's disease as an ethnically diverse group but a group that was nonetheless solidly middle class.

Often enough, Alzheimer's was described abstractly as a highly democratic disease. For example, the authors of The 36-Hour Day, Nancy Mace and Peter Rabins, asserted that Alzheimer's disease and other "dementing diseases know no social or racial lines: the rich and the poor, the wise and the simple alike become victims."[53] But, as the previous examples suggest, whether dementia was actually democratic or not, when it was represented in more concrete terms either through hypothetical or exemplary cases, people with the disease were almost always identifiable as middle or upper class and never identified as being poor or afflicted with some other social problem. This middle-class bias reflected the desire to heighten the sense of the disease's tragedy by emphasizing the worthiness of its victims; it also reflected the background of the leaders of the Alzheimer's disease movement, who tended to be successful middle-class professionals. Nowhere was this bias clearer than when the financial burdens of the disease were discussed. In testimony before the Senate Special Committee on Aging, a lawyer specializing in financial planning for Alzheimer's disease families argued that "it is not possible . . . unless you are very wealthy or very poor, to manage the system." Unless they were very wealthy, families could expect to spend the bulk of their life savings on health care for the person with Alzheimer's disease until they could qualify for Medicaid, a program designed to help the poor. The result, he argued, was a "remarkable development. Middle-class families, driven by fear and panic, seeking to take advantage of what is essentially a welfare program."[54] This point was made by another witness at these hearings, Samuel Sadin, director of the Institute on Law and Rights of Older Adults at the Brookdale Center on Aging. "We are seeing pauperization as the price of Medicaid eligibility," Sadin argued. "The term 'spend-down' may sound better, but make no mistake—what we are talking about is pauperization in old age, of hard-working American citizens."[55] Senator Heinz, chair of the hearing, summarized this testimony: "What we have learned so far is that it is literally better, if you are predestined to become an Alzheimer's patient, it is better to be very rich or very poor, but do not be a member of the group in between." Sadin responded that "even if you are poor, it is very tough," but Heinz continued with additional points about the problems faced by middle-class married couples, who were put in a particularly bad position.[56] Although Alzheimer's disease might have theoretically affected the rich and the poor equally, as a dread disease Alzheimer's was represented in public policy discourse as important because it ruined the lives of decent, respectable people from the middle class. Thus, an important aspect of the discourse of the "health politics of anguish" was its propensity to reinforce the American mythos regarding a diverse but universal and cohesive middle class of decent people who share a common set of values and expectations about the meaning of health and happiness.

"LIKE A FUNERAL THAT NEVER ENDS": THE STEREOTYPE OF THE ALZHEIMER'S DISEASE VICTIM AND THE STIGMA OF OLD AGE

In addition to the conscious aim of reinforcing the central formulation of the Alzheimer's disease movement, representations of disease victims can be read as an attempt to contain anxieties about the fragility of the aging self. The anguish of

Alzheimer's disease was not simply caused by the pain and humiliation it added to a life. All chronic diseases did as much. Alzheimer's disease was the worst of all diseases for what it took away—memory and language, the skills necessary for the maintenance of selfhood. These losses undermined the opportunities for self-expression, personal enrichment, and fulfillment that had increasingly been the means of middle-class self-fashioning in the individualistic consumer culture that had developed in twentieth-century America. Stereotypical representations of Alzheimer's disease were a way of managing anxiety about "losing it," about failing to maintain a stable and coherent self.

As discussed in chapters 1 and 3, stereotypes of senility have been a particularly intense locus of this anxiety in twentieth-century American culture. Stereotypes divide the world into stark images of good and bad, thus allowing us to maintain a reassuring distinction between the two in situations in which ambiguity threatens to overwhelm us. By focusing on the manifest failures of the senile old man or woman to keep their lives together, anxious Americans could reassure themselves that their own selfhood was secure—at least while they were young. Alzheimer's disease advocates believed that the new conceptualization of the disease, based on advances in neuroscience, would lessen the stigma of late-life dementia. But because of the burden of anxiety about the coherence of the self which representations of dementia carried, the stigma of Alzheimer's disease was certainly not less and may in fact have been greater than that associated with the vague concept of senility.

"There is no reason to be ashamed or embarrassed because a family member has a dementing illness," Mace and Rabins assured readers of *The 36-Hour Day.* "Many brilliant and famous people have suffered from dementing illnesses. Although dementias associated with the final stage of syphilis were common in the past, this is very rare today."[57] Similarly, they thought that if the public understood that the unsettling behavior of people with dementia was a symptom of a *disease,* stigma would be lessened. "It is important for those around him to remember that many of the person's behaviors are beyond his control: for example, he may not be able to keep his anger in check or to stop pacing the floor. The changes that occur are not the result of an unpleasant personality grown old; they are the result of damage to the brain and are usually beyond the control of the patient."[58]

But stigma is more than the degree to which the suffering of people with disease can be attributed to something being "wrong with them" or to which they "brought the suffering on themselves." Stigma is the amount of anxiety surrounding the boundary between the normal and the pathological. Put another way, stigma is directly related to the social stakes of a particular set of behaviors or

symptoms that are judged to deviate from some notion of normal. (These behaviors and symptoms may or may not be brought together under the rubric of a
disease category.) The behaviors and symptoms commonly brought together in
the category of "having a cold" are not deeply stigmatized, even though people
who cough, sniffle, and sneeze clearly have something "wrong" with them and
are often "blamed" for their failure to rest properly, take vitamin C, suck zinc
lozenges, or drink echinacea tea. There is simply too little at stake in whether one
does or does not have a cold for it to be highly stigmatized. Throughout the
twentieth century, the stakes involved in behaviors and symptoms variously called
senility or Alzheimer's disease have become enormous—calling into question the
very personhood of those who exhibit them. Alzheimer's attacked the cognitive
skills necessary both to maintain an inner sense of selfhood and, perhaps more
important, to present a stable and coherent self to others. "This illness strikes at
the very core of our being, depriving the individual of the qualities that endeared
them to all around them," wrote one physician in the foreword to a caregiver's
account of the disease. "There is little physical pain, disfigurement, or mutila-
tion. . . . Instead, the disease insidiously robs the victims of their unique thought
processes, their insights, their judgment, their ability to learn new information.
Without these capabilities, the adult human regresses to an earlier, dependent
life."[59] As a result of these losses, people developing dementia were deeply stigmatized—not for having the disease per se but for their resulting inability to carry
off their role as a respectable middle-class person.[60] Nor could attributing these
failures to changes in the brain which were beyond people's control lessen the
stigma, for the loss of self-control, awareness, and personal responsibility was
itself perhaps the most horrifying thing imaginable to middle-class Americans.
Consider the promotional copy on the back cover of a sensational "supermarket"
paperback on Alzheimer's disease entitled The Living Death: "They steal. They
shoplift. They're violent. They 'expose' themselves in public. They're verbally
abusive. They lie. And they don't know any better. Meet some of the 4 million
Americans who have Alzheimer's disease in the pages of The Living Death."[61]
The Alzheimer's disease victim, it seemed, was guilty of the most loathsome
violations of propriety. In his or her total loss of self-control, the stereotype of the
Alzheimer's disease victim was the epitome of the failed individual, just as the
figure of the senile man had been in the nineteenth century.

Representations of the person with Alzheimer's disease were a means of maintaining the distinction between decent, middle-class citizens and the sort of failures invoked on the back of the paperback. Thus, if Alzheimer's disease advo-

cates represented the exemplary Alzheimer's disease victim as a solid, respected middle-class citizen before the disease struck, he or she became someone or something quite different as the disease progressed. The stereotypical Alzheimer's disease victim was in fact at the outer limit of stigmatization—often represented as a stranger, a ghost or shadow, a nonperson. This distinction between the stable, respectable self before the disease and the confused, discredited self that was lost to the disease was at the core of representations of the Alzheimer's victim.

The Alzheimer's Disease and Related Disorders Association instantiated this representation in its publicity materials. For example, in 1984 it began using a graphic that consisted of a series of close-up photos of an attractive, intelligent-looking woman in her mid-fifties. Each of the photos was identical except that, moving from left to right, they faded from sharp black and white to gray, the last image virtually a blank—suggesting visually the gradual effacement of the self. In a brief caption explaining the graphic in its newsletter, the association noted that "the image of the AD victim 'getting further away' is reflected throughout ADRDA's posters, magazines and newspaper advertisements, as well as in the theme song, 'It's a Long Goodbye.'"[62] In a phrase that became common in representations of Alzheimer's disease, Bobbi Glaze, a founding member of the ADRDA, testified that the disease was "like a funeral that never ends." Telling of her own experience, she noted that her husband had been a "handsome, vital, athletic man, a civic leader, a public speaker, a highly respected businessman. He was administrative vice president of his company." But now he was permanently hospitalized, hopelessly dependent, unable for the past four years even to speak. "I have a husband, but I speak of him in the past tense. I am not a divorcee; I am not a widow; but where do I fit?"[63] The idea that the Alzheimer's victim was in some fundamental way already gone despite the unnerving persistence of the body was an almost ubiquitous feature of representations of the disease both within the Alzheimer's disease movement and in the popular press. "What has happened to her humanity? . . . her soul?" a caregiver writing in the *ADRDA Newsletter* wondered of her mother-in-law. "She exists in oblivion but has an amazing tenacity for life. A mere shell that breathes, she 'lives' on and on."[64] Similarly, a teenaged girl who was donating part of her bat mitzvah money to the Alzheimer's Association wrote of her grandfather that "when he went, he wasn't himself. He had been possessed by a ghost, this ghost being the disease of Alzheimer's. For the last three years of his life my grandfather wasn't the man my mother wants me to remember as her father."[65] Another teenaged granddaughter writing in the newsletter expressed such feelings in verse:

No one knows

The hurt I feel

So Deep inside my heart

The pain of

Someone gone away

Not dead

Yet gone forever. . . .

The emptiness and

Pain changes to

Bitterness

Toward the intruder

Who took her place.[66]

In a story on Alzheimer's disease in *Modern Maturity*, the wife of a person with the disease expressed the feeling that an "enigmatic and insidious monster" had taken over her husband's brain. "I look at him and wonder who exactly it is asleep in that chair opposite me. It certainly isn't the man I married. I've become ambivalent about my dear mate because Alzheimer's disease is slowly but treacherously consuming him."[67]

The perverse capacity of the body to outlive the mind, haunting those who had loved the person with Alzheimer's, was often cited as the worst aspect of the disease. Lewis Thomas invoked the "old man's friend" as the best that the Alzheimer's disease victim could hope for: "It is the worst of all diseases, not just for what it does to the patient, but for its devastating effects on families and friends. It begins with the loss of learned skills—arithmetic and typing, for instance—and progresses inexorably to a total shutting down of the mind. It is not in itself lethal, unmercifully; patients go on and on living, essentially brainless but otherwise healthy, into advanced age, unless lucky enough to be saved by pneumonia."[68] *Newsweek* argued that of all the incurable diseases, "Alzheimer's may be the cruelest, because it kills its victims twice. In Alzheimer's, the mind dies first: names, dates, places—the interior scrapbook of an entire life—fade into mists of non-recognition. The simplest tasks . . . become insurmountable. Then, the body dies. No longer able to walk or control elemental functions, the victim lies curled in a fetal position, gradually sinking into coma and death."[69]

Alzheimer's advocates sometimes tried to soften such harsh representations with the assertion that, despite the severity of Alzheimer's disease, the person who had it somehow retained an essential humanity. But this assertion stood in an uneasy opposition to graphic representations of dissolution. For example, throughout another popular caregivers advice book, *The Loss of Self*, there is an unresolved tension between the assertion that Alzheimer's disease is the worst of all diseases because it entails the loss of selfhood and the authors' insistence that the person with dementia nonetheless retained an essential humanity. "Loss of sight, hearing, an arm, or a leg challenges a person to cope with significant change. However, the victim of Alzheimer's disease must eventually come to terms with a far more frightening prospect—the complete loss of self," the authors wrote in the first chapter of the book. "And for the family, according to one daughter, 'the death of the mind is the worst death imaginable.' Family members share a life of emotional turmoil as they witness the disintegration of someone they love." But the authors then assert that "a great deal can be done to ease the burden of Alzheimer's disease and related disorders." After the diagnosis, "an individual may live five, ten, or fifteen years or more. These are long human years. . . . If the patient and family are able to prepare themselves to deal with the future, there is time for them to live and love, despite the ravages of a progressive brain disease." Although they were no doubt sincere that "our patients and families have taught us that there is life that transcends Alzheimer's disease," it is difficult to reconcile this lesson with the idea that "both the victim and family suffer with the inexorable dissolution of self."[70]

The unresolved tension between the assertion that the person with Alzheimer's disease loses selfhood and yet retains an essential humanity is also present when the authors address the issue of stigma directly. Stigma, they argued, was "one of the most significant barriers to providing good care." They recognized that the tendency to treat the person diagnosed with Alzheimer's disease differently was understandable: "Cognitive deficits, communication difficulties, and strange behaviors make most of us frightened. These feelings color our abilities to see and react to the 'humanity' of the patient." Placing the word *humanity* in quotation marks was a curious authorial gesture when their aim was to prevent the stigmatization of the patient. Perhaps it was intended as emphasis, but it was more easily read as a suggestion that the humanity of the person with Alzheimer's disease was in fact provisional or ambiguous. This ambivalence was evident in the next paragraph as well, when the authors urged readers "to identify and overcome false beliefs and fears about dementia" in order to overcome the problem of stigma: "While Alzheimer's disease involves the slow, irrevocable loss of ability and ultimately even the sense of self, the stereotypical notion that victims of dementia lose overnight the ability to think, write, read, talk, work, or love is a tragic error."[71] The first part of the sentence invokes the most frightening aspects of Alzheimer's disease, while the second part is intended to soften this by suggesting that these losses may occur gradually. But the slow pace of deterioration was cold comfort at best and, indeed, was often seen as particularly torturous—the slow accumulation of unbearable suffering and humiliation. The total loss of self was really just a matter of time, which was perhaps the meaning of bracketing the word *humanity* with quotation marks in the previous paragraph. *The 36-Hour Day* quoted a caregiver who expressed this feeling: "Sometimes I wish he would die so it would be over. It seems as if he is dying a bit at a time, day after day. . . . It seems like I'm on an emotional treadmill going around and around and it's slowly wearing me down."[72]

Caregivers and family members sometimes tried to resolve the contradiction between understanding the person with Alzheimer's disease as "empty shell" and as essentially still human by describing moments when the "real person" briefly reappeared, only to sink back into the void of dementia. In *The Loss of Self*, a woman describes such a moment with her husband, who had dementia and whom she was caring for at home but with whom she could scarcely communicate. He had been an avid bird-watcher throughout his life, and although his binoculars, his books, and his sketches of birds were meaningless to him now, he continued to enjoy watching birds at the backyard feeder. One day a finch flew into the glass door of their house. "I walked into the living room, saw the bird on the porch, and saw Robin [her husband]—crying." After determining that the bird was dead, they buried it in the backyard together. "Robin held a brown paper bag while I dropped the bird in. He folded the top down neatly several times and gave it back to me. . . . An hour later I know he did not remember what happened. But for a few minutes that day I had a glimpse of the gentle man who loved the world of nature. Somewhere deep inside, part of him was still there, but I couldn't hold onto it."[73] Writing in the *ADRDA Newsletter*, a teenaged girl employed an elaborate metaphor to describe the loss and fleeting return of her grandfather's selfhood. "Alzheimer's disease, a thief who returned night after night to the same home, slowly turned my grandfather into an abandoned house. . . . His eyes, gaping windows, are void of light. . . . The huge old rooms of his soul echo when one cries out in them. . . . The inhabitants, his hope and spirit, deserted him long ago." During a visit with her grandfather, she examined a music box that she had long associated with him:

When I lifted the lid and heard the familiar music, he looked up and said with an expression of reminiscence, "We got that music box from Midori. She was our foreign exchange student from Japan 25 years ago . . . we never did hear from her again, did we?" I glanced up with surprise, to see him already retreating into his empty house. The fact that after being able to grasp nearly nothing around him, my grandfather reached out to the one thing which was the epitome of my memories of him as he once was, is the last thing my grandfather was able to give me. I do not believe I will ever truly see him again. . . . My grandfather is an abandoned house.[74]

Such moments were usually seen in this positive light, as authentic and poignant respite from the inexorable losses of dementia. But they could also be seen as cruel deceptions that only made dealing with dementia more frustrating. "Having a parent with Alzheimer's fools you into believing there's a real person there in front of you—instead of the shadow that's really there," a daughter said of her mother in the paperback *The Living Death*. "It's hard to deal with her, because some of the time she's the same wonderful person I've known for over 30 years, and at other times she's somebody else. You get tricked into believing it's the same person, even when she displays this weird behavior."[75] Whether such moments were interpreted positively or negatively in caregiver accounts, they did not challenge the basic stereotype of people with Alzheimer's disease as losing their selfhood. They were not seen as signs of an intact self struggling with the tremendous challenges of dementia but as remnants—usually comforting but sometimes disturbing—of the self that had been destroyed by the disease. The basic distinction remained between the "real" person, possessed of a stable and coherent self before the disease, and the "stranger" or "shell" that the person became as the dementia progressed.

Although the stereotype of the Alzheimer's disease victim bore much the same burden of anxiety about the fragility of the self as the earlier stereotype of the senile, it did differ in one significant way. Earlier stereotypes of senility maintained a boundary between the pathological decadence that was old age and the rest of the life course, particularly youth and middle age. The stereotype of people with Alzheimer's disease sought to move that boundary into old age itself, distinguishing between a particular disease condition and normal old age, which was free of such catastrophic losses. Representing the person with dementia as the victim of dread disease theoretically preserved old age as a stage of life relatively free of deterioration and dependency. The person with dementia continued to figure as the epitome of the failed self but was now contained within the sharper, more restricted boundaries of a particular disease category. But maintaining the boundary between old age and the pathological processes of dementia was difficult. The distinction between Alzheimer's disease and normal aging was the central dogma of most scientists working on Alzheimer's disease, but the issue remained problematic in the scientific literature. Conclusive evidence that Alzheimer's disease was distinct from aging simply did not exist. In popular and political discourse, Alzheimer's advocates even more vehemently asserted that Alzheimer's was a disease and not aging. Nonetheless, there was a constant tendency to conflate the two because Alzheimer's disease was so clearly an age-related disease. This was highlighted by the "apocalyptic demography" employed in trying to win funds for the disease, as could be seen in Lewis Thomas's essay on dementia. On the one hand, Thomas asserts that senile dementia "is not, as we used to think, simply an aspect of aging or a natural part of the human condition, nor is it due to hardening of the arteries or anything else we know about. It remains an unsolved mystery." Yet elsewhere in the essay Thomas justifies calling "senility, or, as it is now termed, senile dementia" the "disease-of-the-century" by noting that it "afflicts increasing members of our population because of the increasing population of older people in the society." Alzheimer's disease, the major form of the disorder, "affects more than 500,000 people over the age of fifty, most of them in their seventies and eighties," who fill the country's nursing homes at a cost of tens of billions of dollars per year.[76]

The conflation of aging and Alzheimer's disease could also be seen in efforts to relieve anxiety. An NIA publication designed to refute the "myth of senility" and reassure people that minor memory lapses did not necessarily signal the onset of dementia began by stating that " 'senility'—something which most of us fear in old age—is not a normal sign of growing old; in fact, it is not even a disease." It goes on to argue that senility is a commonly used word that describes cognitive, behavioral, and personality changes and that "the important thing to remember is that some of the problems which are generally referred to under the medical description of senile dementia can be treated and cured, while others, at this time, can only be treated without hope of restoring lost brain function." Regarding the latter, Alzheimer's disease, in "which changes in the nerve cells of the outer layer of the brain result in the death of a large number of cells," is the most significant of these.[77] The lesson being taught here was not that there is no such thing as senility but that, in words of another popular layperson's guide, *The Myth of Senility*, Alzheimer's disease was the "real senile dementia." While assertions that only 20 percent of people over 65 are afflicted with Alzheimer's disease are meant to be comforting, this is offset by the scientific precision and certainty with which the neurological damage and consequent dementia are described. As Robert Butler put it in the foreword to the book, "for as many as three million elderly Americans, 'senility' is no myth."[78]

The ease with which Alzheimer's disease could be conflated with aging was perhaps most evident when health hucksters used the fear of Alzheimer's disease

to hawk antiaging products. A slick brochure from Gero Vita Laboratories announced that there was a "Senility Epidemic—1 in 5 over 60 Affected" and invited
readers to look inside to read about "The Causes and How to Deter It." The
brochure pitched a product called GH3 by linking "age spots" on the skin with
Alzheimer's disease. The first page of the brochure featured a micrograph of
neurons, much like those that frequently appeared in news articles on Alzheimer's disease, with a headline warning that "Age spots on your skin signal that a
brown slime is forming on the neurons of your brain!" The text explained that age
spots are technically known as "lipofuscin" and that lipofuscin in the brain forms
a brown slime on the neurons. "As the slime thickens, senility and dementia
increase." The micrograph showed "a section of neurons in the brain of a man
who died from a stroke. He was extremely senile. As you can see, the brown slime
covered every neuron." The next page featured a mock news story attributed to
"Marcus Wellbourne, senior science editor," under the headline "If You Have Age
Spots . . . Don't Wait until Your Memory Gets Worse." It featured a color photo
of a bright-eyed older Caucasian woman whose face had a number of brown
blotches. The caption repeated the explanation about lipofuscin, age spots, and
neuronal slime, claiming that "as you develop age spots on your skin . . . short-
term memory begins deteriorating." It also included a quote from the *Los Angeles
Times* concerning the increasing prevalence of Alzheimer's disease. The text
played explicitly on the fear of Alzheimer's disease: "Probably the scariest aspect
of getting older is the possibility that we will become senile with its attendant
forgetfulness, mental confusion, inability to manage our money and vulnerability
to being taken advantage of. Worse yet, senility is the first step toward various age-
related dementia [*sic*], including the awful Alzheimer's disease." The text went on
to claim that "age spots or lipofuscin are well-accepted in medical science as the
first warning that we are progressing in the direction of senility or dementia."[79]
These claims were, of course, deceptive, to say the least. While the formation of
lipofuscin in the brain had long been noted by scientists, it had never been linked
with mental deterioration. These claims indicate that unscrupulous marketers
thought that the widespread fear of Alzheimer's disease could be conflated with
the fear of aging to sell their products.

If in the discourse of Alzheimer's disease advocacy there was a greater awareness than in earlier decades that severe mental deterioration was not an inevitable
nor even common occurrence in old age, there was nonetheless at least as much
fear about the prospect of Alzheimer's disease as there had been about the vaguer
concept of senility. Alzheimer's disease generated such intense fear in part *because* many people could now expect an old age that was relatively free of physical,

social, and economic deprivation. Part of the tragedy of Alzheimer's disease was that it meant the loss of opportunities for self expression, personal enrichment, and fulfillment that had become an expected part of late life in the prosperous decades after World War II. Thus, representations of people with Alzheimer's disease often begin with a lament for the loss of the well-deserved and longed-for leisure of retirement. "My mother had looked forward to her retirement, having worked almost all of her life as a telephone operator beginning when she was a teen-ager, as a housewife for a few years until my father died when I was 2, as a secretary for the next 3 1/2 decades," wrote *Chicago Tribune* journalist Charles Leroux of his mother's dementia in one of the earliest major newspaper accounts of Alzheimer's disease. "The little jokes she enjoyed; the financial independence she had won after nearly a half century of work, the joy she found in minor lovelinesses of life . . . all have become victims of Alzheimer's disease. Just when she was getting ready finally to enjoy life, the capacity for enjoyment was taken from her. It wasn't fair." Leroux quoted family members at a support group meeting in Chicago which he thought "could have been any of the hundreds of Alzheimer's support groups meeting all over the country," who described Alzheimer's disease the same way. " 'They feel something has been snatched from them,' one of the women said about her parents' hoped for retirement years. Her father had intended to sell the business he had built, but now he has the beginning of Alzheimer's and can't handle complex dealings leading to a sale; and instead he's trying to continue the business and in his failing grasp it's falling apart and becoming less salable every day."[80] A caregiver writing in the *ADRDA Newsletter* expressed a similar sense of loss over her 62-year-old mother-in-law's Alzheimer's. "They had saved and invested their money well, and had ample means to travel around the world, or do anything else they wished. They were finally free to reap the rewards of life and to delight in their two grandchildren," she wrote. "They were people who you might call 'the salt of the earth,' for all their lives they had given and done for others. So why this cruel twist of fate, now? Why, anytime? She was still young. And, so was he."[81] Another caregiver writing in the newsletter about her husband's dementia expressed this sentiment in verse:

You dream of the day when you can retire
And do all the things that you desire
But just when you're getting ready to start
You get some news that just breaks your heart
Your spouse was diagnosed—Alzheimer's Disease
Not him you say, Oh no God Please![82]

By the 1970s, if the prospect of retirement had become attractive to most older Americans, Alzheimer's disease figured as perhaps the ultimate threat to that prospect; if "senility" was now relegated to the status of myth, older people now had to contend with the prospect of a very real and thoroughly devastating disease that could rob them not only of their retirement but the very essence of their lives.

The intense fear that Alzheimer's aroused was often a part of representations of Alzheimer's disease. "This diagnosis really upset me. You see, I had read about Alzheimer's disease and I had some idea how awful it could be," a 61-year-old woman said of her reaction when her depression was misdiagnosed as Alzheimer's disease. "I thought as I left the doctor's office that day that I probably should sell my house. Fortunately, my sister talked me out of that, but I was very depressed. I worried that I wouldn't be able to work any more and that my life would gradually fall apart."[83] *The Loss of Self* quoted a woman who, despite being aware of cognitive problems, put off seeing a doctor for several years because of the fear that Alzheimer's disease aroused. "My memory lapses were humiliating. I couldn't depend on myself anymore. Mysteriously I was changing. I knew about this Alzheimer's disease, how it destroys intelligent behavior in bits and pieces, but I was afraid to think it was happening to me." When she finally was given the diagnosis, she felt a sense of utter helplessness. "The very words sounded harsh and unreal. I wanted to believe that he was talking about someone else—not me."[84]

The "health politics of anguish" described in this chapter was tremendously successful at generating money for biomedical research, but it is still too soon to tell whether that success will result in a decisive medical intervention. In any case, this strategy has been a failure at winning support for caregivers. But Alzheimer's disease advocates had faith that medical science brought other benefits as well. Knowledge dispelled ignorance, and it was ignorance that created fear and hostility toward old age. As Alzheimer's disease advocates worked to educate the public about the "true nature" of aging and dementia, they believed they would inevitably lessen the stigma attached to both old age and dementia. Dementia was a disease, and a disease that, though more prevalent among older people, did not inevitably afflict them. Knowing this, it was thought, would relieve most older people of needless concern that they would lose their mind as they aged, while generating greater understanding and genuine sympathy for those who did develop dementia. Although discourse on Alzheimer's disease did indicate a change in the meanings of old age and dementia, in some ways it actually heightened anxiety about the prospect of losing one's mind as one grew older. If American culture had produced more positive meanings for old age in the 1980s and

1990s, it was less tolerant than ever of disability and dependence. These developments were thrown into sharp focus in the Janet Adkins case, which marked the
start of Jack Kevorkian's career as a self-described "obitiatrist."

"HER MIND WAS EVERYTHING TO HER":
PHYSICIAN-ASSISTED SUICIDE AS A RATIONAL
RESPONSE TO THE PROBLEM OF DEMENTIA

When Janet Adkins, a 54-year-old former college instructor, was diagnosed
with Alzheimer's disease, she determined that she would end her life before
crossing over into degrading dependency. When in 1990, suffering the minor
impairments of early-stage Alzheimer's, she traveled from Oregon to Michigan to
become the first client whose suicide was facilitated by Dr. Jack Kevorkian, she
made physician-assisted suicide and voluntary euthanasia a national issue. Her
short suicide note, which her husband, Ronald Adkins, and a close friend who
had traveled with them signed as witnesses, reflected a calm determination to end
her life before the ravages of Alzheimer's disease could destroy it. "I have decided
for the following reason to take my own life. This is a decision taken in a normal
state of mind and is fully considered," she wrote, the slight lapses in grammar
and punctuation perhaps indicating deterioration of her cognitive abilities. "I
have alzheimers disease and do not want to let it progress any further. I dont
choose to put my family or myself through the agony of this Terrible disease."[85]

The Adkins case and the career of "Dr. Death" that it launched were part of a
tremendous change in American attitudes toward death in post–World War II
America. In part, this change was a reaction to one of the problems that accompanied modern medicine—the transformation of death from a discrete event to a
protracted, often painful and humiliating process whose boundaries were ambiguous. While a strong belief in the power of modern medical technology to van-
quish disease and suffering persisted in American culture, it was joined by a
rising fear that medical technology gone wild could overwhelm individual autonomy. The specter of being kept nominally alive on a respirator in a hopelessly
dependent or vegetative state for months and perhaps years was by the 1980s as
prominent a representation of modern medicine in American culture as the
images of miracle drugs, heroic surgery, and omniscient diagnostic technology
which had been dominant in the first two-thirds of the century.[86]

These new attitudes toward death were also fueled by the civil rights and
women's liberation movements, which inspired a host of other claims for rights
against what were seen as oppressive forces within society. One such claim was

for patient's rights, which included the right to "death with dignity." Death with dignity was usually thought of as the right to decline life-sustaining medical treatments while receiving effective palliative care, but many people sought to make physician-assisted suicide and voluntary euthanasia available to patients as well. Coupled with this increasing tendency to claim rights was a growing emphasis on personal fulfillment since the 1960s and 1970s, especially with regard to tuning and being in tune with one's body and its natural rhythms. In such a cultural milieu, the prospect of losing control over the last phase of one's life came to seem intolerable, and Americans sought ways of taking control. In the nineteenth century a "good death" meant one in which the person embraced suffering and loss as a preparation for heaven; in the twentieth century a "good death" meant painlessness and even fulfillment. In the culture of consumer capitalism, dying increasingly became another means of self-expression, if not self-indulgence. The desire for control of one's death became another mechanism of denial and suppression. For example, one 1970s book on the *ars moriendi* asked whether "there is a person alive who wouldn't like to go to their dying full of excitement, without fear and without morbidity. This book tells you how." Another book expressed this feeling more succinctly in its title: *To Die with Style!*[87]

This change in the meaning of death is closely connected to the transformation in the social circumstances and cultural meaning of old age described in chapter 3. With changes in public policy that made old age more financially secure and health care more accessible, the prospects for growing older never seemed brighter. All of this was fueled, as with the changing attitude toward death, by an increased emphasis on personal fulfillment and the tendency to claim and defend rights. But, as with "modern dying," there was a dark side to this. It was far from clear that increased longevity would mean freedom from deterioration. Although gerontologists such as Butler might insist that there was a distinction between aging, which was free of significant deterioration, and the disease processes that were associated with aging, the prospect of living longer only to face debilitating illnesses caused significant anxiety, especially in the case of Alzheimer's disease. If Americans had come to expect a much more comfortable and fulfilling old age, the possibility of deterioration and loss was much less tolerable. As aging, like death, became something that Americans expected to control and even utilize as yet another route to personal fulfillment, instances in which control was taken away and suffering had to be endured were intolerable intrusions into the process of self-creation.

The Adkins case was particularly powerful and controversial because it brought into focus all these developments. Even before her diagnosis, both Janet Adkins and her husband, Ronald, had been members of the Hemlock Society, the national organization dedicated to empowering individuals to end the suffering and humiliation of terminal illness through suicide or voluntary euthanasia. Derek Humphrey, the society's founder, has written that suicide is a rational choice not only for those who were terminally ill but also for the elderly as a preemptive alternative to the deterioration they were likely to face as they grew older.[88]

Certainly, all accounts of how Janet Adkins arrived at her decision to end her life with Kevorkian's assistance suggest that these were her motivations and concerns. Ronald Adkins said that his wife had consistently expressed a desire to take her life from the time she was first diagnosed with Alzheimer's disease and was told by the physician making the diagnosis that Ronald would have to care for her completely within a year. Her sons supported her right to make the decision but disagreed with it and convinced her to enroll in a clinical trial of tacrine, which brought neither improvement nor delay in her deterioration. Although the medical team in Seattle that administered the trial told her that the course of her deterioration would be much more gradual than the initial doctor had indicated and that she had at least two or three years, and possibly much longer, of relatively high quality life before reaching the advanced stages of the disease, within a few months she made arrangements to travel to Michigan for Kevorkian's help, arguing that she needed to kill herself before the disease progressed to the point where she would be judged incompetent to decide her own fate.[89] "This was a very special woman," Ronald Adkins told reporters after her death. "She loved life. She was upset because she was losing her mind and her mind was everything to her."[90] Her sons, friends, and the minister with whom she planned her upbeat funeral service all confirmed that she knew exactly what she was doing and that she "approached her death with the same zest and independence that she had shown during her life."[91] The picture of Adkins that emerged in newspaper accounts was the epitome of "successful aging"—a woman who embarked on the passage through middle age with confidence and vigor, taking on new challenges and new avenues for fulfillment. But she was also the epitome of someone keenly aware of the thin line between the pathological and the normal, between the mentally sound and the demented—a woman who could not countenance the prospect of "losing it." As one journalist put it, "Janet Adkins knew what she wanted. She had played tennis, climbed the Himalayas, traveled around the world, hang glided, played Brahms duets with her husband, raised three sons, taught piano, taught English, learned T'ai Chi, studied reincarnation. Being addled and dependent was not her idea of living."[92]

Not surprisingly given Kevorkian's desire for publicity, the Adkins case drew a tremendous amount of attention, and bioethicists were called on to make sense of it in accounts in the popular media. Although Kevorkian's actions were all but universally condemned, few bioethicists found grounds to criticize Janet Adkins' decision to end her life. "It is a Catch-22 situation," an expert on health law, quoted for an article in the *New York Times*, said. "You want to hang on as long as you are competent, but after you're incompetent and life is not worth living, it's too late, you can't do it anymore."[93]

Somewhat paradoxically, Kevorkian's critics objected to his decision to assist Adkins' suicide both because she enjoyed a relatively high quality of life up to the time she pushed the button on his homemade suicide machine *and* because the diagnosis of Alzheimer's disease brought into question her competence to make the decision. On the one hand, virtually every report of her suicide in newspapers and the weekly newsmagazines emphasized her vigor and capacity for enjoyment of life by mentioning that only the week before she had beaten her 32-year-old son in a tennis match and in her last few days had been able to enjoy a romantic weekend with her husband. On the other hand, most of these accounts also raised the question of whether she was competent to make her decision. For example, *Time* noted that "one significant symptom [of Alzheimer's disease] is sufficient mental deterioration to impair the ability to make decisions," and it quoted an expert who noted that Kevorkian needed to claim that she made her decision competently but that "the diagnosis of Alzheimer's is almost incompatible with that claim."[94]

Robert Butler came the closest to criticizing Adkins' decision, expressing concern about the impact of her suicide on the morale of the elderly: "It's very demoralizing to hear of a 54-year-old giving up life when you're in your 80s and have heart disease and arthritis and some dementia and are still surviving, maybe working and taking care of your spouse."[95] This quotation was unusual, however. More typically, commentators saw her either as a victim of Kevorkian's irresponsible behavior or as someone dealing rationally, if not heroically, with her situation. An editorial in the *New York Times* compared her suicide to a death-row execution. "One might . . . say that Mrs. Adkins had a choice whereas the condemned do not. But did she really? Janet Adkins had Alzheimer's disease. Her family was divided. Whether she was competent to make such a decision, as Dr. Kevorkian contends, is questionable. But whether Kevorkian is competent to decide on *her* behalf is not even questionable."[96] On the other hand, the editor of the *New England Journal of Medicine*, Marcia Angell, wrote in an op-ed piece that patients with Alzheimer's disease can expect their brain to be destroyed slowly over many years, during which they become increasingly dependent on and burdensome to their families. From this perspective, the decision to end life seemed rational. "The prospect for Janet Adkins was bleak. Moreover, modern medical care permits longer and longer survival under these circumstances, and patients are often subjected to the full panoply of aggressive treatment simply because it is available. What is someone like Janet Adkins, who valued her independence, to do?"[97]

Recognizing this predicament, most critics focused not on Adkins' decision to end her life but on Kevorkian's decision to help. Angell found this position puzzling: "If suicide under these circumstances is not wrong, then why is it wrong to help?" Angell went on to argue that if "a doctor's first obligation is to help the patient, consonant with the patient's autonomy," then doctors sometimes had to take actions that shortened their patient's life, as when the dosage necessary to manage the pain of terminally ill patients put them at risk of succumbing to secondary illnesses such as pneumonia. Most of Kevorkian's many critics did not focus on whether it was wrong for Kevorkian to help but on his method of doing so—the fact that he had met the patient only one time before helping with her suicide and that he had not even employed the procedural safeguards called for by right-to-die groups such as the Hemlock Society.[98] Criticism even centered on the aesthetics of her suicide. Virtually all the news accounts of the event emphasized that her death had occurred in the back of Kevorkian's rusty 1968 Volkswagen van, which he'd had to use because there were no hotels or other facilities willing to rent him space for the purpose. On this score, even Hemlock Society founder Derek Humphrey was critical of what occurred: "It's not death with dignity to have to travel 2,000 miles from home and die in the back of a camper."[99] But, given the ravages of Alzheimer's disease, the Janet Adkins' decision to end her life was hardly questioned.

Adkins' decision and Kevorkian's defense of his action are steeped in what Stephen Post has called the "hypercognitive values" of our society: self-control, independence, economic productivity, and cognitive enhancement.[100] In the book he wrote a year after Adkins' death, Kevorkian defended his decision to help Adkins with her suicide in unabashedly "hypercognitive" terms. Although he was well aware that the decision to accept her as his first client would be subject to the "criticism of picayune and overly emotional critics," he viewed her as "a qualified, justifiable candidate if not 'ideal.'" Kevorkian acknowledged that "from a physical standpoint Janet was not imminently terminal" but that "there seemed little doubt that mentally she was—and, after all, it is one's mental status that determines the essence of one's existence."[101] To the extent that this emphasis on cognition remained the essential element of the ideal of human fulfillment, preemptive assisted suicide and voluntary euthanasia (killing oneself or having oneself killed before dementia produces severe impairments) will seem justified as a rational solution to the problem of dementia.[102]

Thus to some people Adkins might seem not a victim but, in the words of one bioethicist, a "heroine for rational suicide."[103] But if she was a hero for the cause of physician-assisted suicide, she was an unwitting one, for according to her husband, neither of them had foreseen the possibility that her case would attract worldwide attention.[104] Moreover, however calmly and rationally she acted to bring about her death, she did so on the basis of an underlying fear. As Kevorkian explained it to reporters: "She was very calm. She dreaded what would have come. I would too. I don't want to die of Alzheimer's—smeared with your own urine and feces, don't know who you are. Come on!"[105]

CHAPTER SIX

The Preservation of Selfhood in the Culture of Dementia

In Stephen Post's critique of the idea, dominant in bioethics, that the cognitive losses of dementia make it unnecessary and perhaps impossible to regard people with dementia as fully possessed of the moral status of personhood, he suggests that fellow bioethicists who espouse such views are practically irrelevant: "Nobody would approach a family in the throes of medical decision making about their loved one with advanced dementia and proclaim that he or she is a 'nonperson' (the rough equivalent of 'shell' or 'husk')."[1] One would hope that Post is right that no clinical bioethicist would push the point on a family at a crucial moment. Yet in a larger sense, of course, those family members have been told these very things again and again, for the "loss of self" that the disease supposedly entails has been a dominant trope of the Alzheimer's disease movement and the "health politics of anguish" described in the previous chapter. It is hard to imagine anyone dealing with dementia, whether as a caregiver, a family member, or a person diagnosed with dementia, who has not participated in or listened to a discussion of the idea that Alzheimer's disease destroys the selfhood of its victims.[2]

Throughout this book, I have explored how this idea is embedded in American social and cultural history. In modern America, the emphasis on self-control through strength of will, self-creation through personal effort, and self-fulfillment through the creative and passionate pursuit of one's desires puts the social and moral status of the cognitively impaired in doubt. When people with dementia lose the ability to articulate a coherent narrative that connects their present with the past, to turn their desires into a rational plan that connects the present to the future, and ultimately to control their most basic physical impulses, their selfhood seems to evaporate in the suffocating atmosphere of an eternally desolate now. In the face of such catastrophic losses, it seems an affront to the dignity of all that an individual was to claim that he or she continues to be the same person—or a person at all—in advanced dementia.

As witnesses to these terrible losses, we have no trouble recalling things. We can cite many incidents that marked the trajectory of this deterioration. We *know* that the person with dementia is the same person, yet we find it excruciatingly difficult to construct a narrative that connects the demented person we see before us with the "self-possessed" person we knew. Quite literally, we have trouble "remembering" or fitting together the demented person with his or her past. And so we speak—however nonsensically—of the person with dementia as being someone or something else, as no longer really there despite the troublesome persistence of the body. In this way, Alzheimer's disease can be seen to create a disruption of memory for both its primary victims and those around them who persist in being called "normal." In its disruption of our ability to keep the past, present, and future in a coherent relationship, the disease impairs us all.

In *The Wounded Storyteller*, Arthur W. Frank argues that all serious illness is experienced as this sort of disruption of memory, a dislocation of the relationship between past, present, and future: "The present is not what the past was supposed to lead to, and whatever future will follow this present is contingent."[3] But this disruption of memory, this temporal disconnect and its attendant disorientation, characterizes not only illness but the totality of experience in what Frank calls "postmodern times."[4] Many scholars have argued that the United States and Western Europe have been undergoing for the past several decades a transition from modernity to some other form of social and cultural organization, variously called postmodernity, high modernity, or late capitalism. Although these various ways of characterizing this transition differ in important ways—most fundamentally on whether they are framed as description or critique, whether this transition is to be lamented or celebrated, encouraged or resisted—these scholars agree that modern notions of the coherence and rationality of time, space, and selfhood are besieged by disorientation and skepticism. In this context, dementia can be seen as one of the emblematic diseases of our times, just as hysteria was in the Victorian era.[5] Alzheimer's disease is deeply troubling both because of what it does to people who have it and because people with this disease remind us so pointedly of our own difficulties in creating and maintaining selfhood. We need to find ways to listen to and appreciate the experience of people with dementia not only, as bioethicists such as Stephen Post argue, to create a truly ethical approach to our treatment *of* them but because we need to bear witness to their struggle as a shared experience.

In this concluding chapter, I discuss the literature produced by professional and family caregivers and people with dementia themselves who have struggled to see people with Alzheimer's as fully human despite the devastating losses that the disease entails and the stigmatizing and dehumanizing terms in which the disease has been represented in popular and professional discourse. This effort has come primarily from two directions: the perspective of religious belief or spirituality, and the perspectives of various strands of social constructionist theory. This is not to say that efforts at challenging the notion that the self is lost are limited to these two broad groups. Sensitive clinicians have probably always been supportive of the selfhood of their patients. One such clinician, Joseph M. Foley, wrote that "it is important to identify functions that are lost, but even more so to identify functions that are preserved. . . . In the formulation of public and institutional policy we must beware of simplifications and generalizations; we must recognize that individual demented persons have their own unique attributes and that, despite metaphors loosely thrown around, they each remain a person, with their own gratifications and frustrations, their own unique background, and their own unique destiny."[6] Mainstream researchers in medical sociology and gerontology have also been paying more attention to the perspective of the person with dementia in addressing issues such as quality of life, treatment efforts, and management of psychosocial complications.[7] However, it is primarily from the perspectives of religious belief and social constructionist theory that challenges to the notion that the person with dementia becomes a hollow shell have been articulated. Moreover, these discourses also speak more directly to the broader cultural issues of memory and dementia which this book explores.

"THAT PART OF HIM REMAINED UNTOUCHED": RELIGIOUS BELIEF AND THE SELFHOOD OF THE PERSON WITH DEMENTIA

Not everyone accepted the idea, dominant in Alzheimer's disease discourse, that selfhood was completely obliterated in dementia. Though perhaps this was hard to detect because the stereotype of the Alzheimer's victim as the lost self was one of the central features of a massive media campaign to win support for Alzheimer's disease, many caregivers and family members insisted that their loved ones continued to retain an essential humanity and connection with their former lives. A small number of academic researchers, bioethicists, and clinical practitioners joined them, producing a body of work that, from a variety of theoretical perspectives, rejected the notion that selfhood was destroyed in dementia and at times even suggested that the boundary between dementia and normality was arbitrary.

That there were brief respites in the inevitable downward spiral of dementia, moments in which a glimpse of the "real person" could be seen, was a commonplace in caregiver accounts of dementia. Often, these moments were used in caregiver narratives to signal closure, a final farewell. But, as shown in Jaber Gubrium's ethnographic studies of caregiver support groups, some caregivers challenged outright and forcefully the notion that the self of the person with Alzheimer's disease was ever truly lost. For example, Gubrium describes how one caregiver support group articulated a "folk" theory of dementia that rejected the notion that the self is lost with cognitive skills in favor of a view that an essential self remains intact, a self that can be reached by using nonverbal, noncognitive means of communication. After a newcomer to the group, in describing a frustrating incident with her husband (who had dementia), suggested that "to me, he's just gone, or getting there fast," veteran caregivers in the group gently chided her for this attitude and provided evidence from their own experience which contradicts the dominant notion that the person with dementia is no longer really there. "If you can't communicate the usual way, I say try another way. Words aren't everything you know," one of the women concluded. "Empty shell? Don't believe it. What do the doctors know? They're right in there somewhere. Maybe they're lost because the brain ain't sparkin' or something, but they're there. You can't give up on them that easily, believe me."[8]

The basis for this assertion that the person with dementia retains his or her selfhood lies in a turn away from the assumption that selfhood is rooted in cognitive abilities, especially language and memory. As Gubrium's ethnographic data show, caregivers who thought that people with dementia remained somehow their "true selves" theorized selfhood as residing in the realm of feeling, a romantic selfhood that could in some ways seem more authentic because it was unencumbered by rationality and culture. Surprisingly, Gubrium does not report that religious belief or spirituality is a factor in whether one believes that people with dementia still possess some essential humanness.[9] While most of the book-length accounts by caregivers equate the loss of cognitive skills with the loss of self, several elaborate more fully this turn toward emotion as the basis of selfhood, and spirituality is in all cases a crucial element.

Writing from the perspective of her Lutheran Christianity, Theresa Strecker described the ways in which the suffering she experienced herself and witnessed in her mother and father as he succumbed to Alzheimer's disease brought her closer to God. Her view of the self in dementia is based on the religious idea of soul or spirit. "We know that there is a living spirit in the Alzheimer's victim, a spirit who still desires the love and compassion of those around him. One has only to visit an Alzheimer's ward to encounter patients who will hold your arm and not let go. These people are not dead. They are alive and need to love others and have others love them." Although the spirit remains intact in dementia, a society that confuses intellect or function with true essence of a self fails to recognize this. The meaning of suffering in Alzheimer's is to remind us of our dependence on each other and on God. "Alzheimer's strips away the characteristics we use to identify one with the world, leaving the spirit of the person intact. It is that spirit that is able to respond to love and knows the comforting presence of another."[10] Although her father is bereft of any memory of things, even of names, he remains connected to his past and to his family at the deeper level of the spirit. Strecker describes this in an episode that began when she returned from a walk with her father, who had severe dementia, and he failed to recognize the house they lived in for thirty years as his own. Only by luring him with the promise of a banana could she get him to enter the house. "This left me with little hope that he remembered anything about our life as a family," she wrote. "I was mistaken." After finishing the banana, he sat on the couch as he did most afternoons, and she and her mother sat down in the kitchen for coffee and a few minutes respite. Soon, he appeared in the doorway with a look of delight on his face:

> He held in his hands a photograph album my mother had made a few weeks before. We stood next to him as he opened it carefully and looked at each picture as if for the very first time. His finger passed over faces, but paused over those of his wife and daughters. "Beautiful child," he said softly, pointing to his grandchild. I pointed to my picture and asked who it was. He studied it carefully and tapped it as he tried to find the answer in his memory. Finally, I told him. He smiled back shyly. My eyes filled with tears as my parents slowly turned the pages of their life. He may have forgotten our names, but not that he loved us. That part of him remained untouched by Alzheimer's.[11]

Ann Davidson's eloquent account of a year in her husband's journey into Alzheimer's disease is more secular and cosmopolitan than that of Strecker. Nonetheless, a more general, one could perhaps even say secular, spirituality rooted in an attraction to Eastern mysticism and a commitment to self-fulfillment is at the core of her ability to recognize the essential humanness of her husband, Julian, despite his deepening dementia. Listening to her husband's tape of Thich Nhat Hanh, the Vietnamese Buddhist monk and popular author of self-help books, as she does her morning walk is the closest she comes to consulting clergy.

Ann Davidson was a speech pathologist and her husband, Julian, a prominent physiologist and sexologist at the Stanford University School of Medicine. In the year that the book described, he had mild to moderate cognitive impairments. Julian's greatest fear was losing the ability to talk, a fear shared by Ann and one that was gradually coming to pass. "I realized that our old easy talk was gone, remarks exchanged while dressing or undressing, while making coffee and putting bagels on the table, while doing dishes or folding laundry. Julian can't talk and do something else at the same time anymore—this man who used to do six things at once." At the same time, she found that to understand his increasingly convoluted formulations, she had to give his speech undivided attention. As his dementia deepened, he increasingly relied on her ability to unscramble his speech—a role she found both frightening and irksome. She wondered what it was about this role that was so terrible. "It's not that one garbled exchange is so difficult," she writes. "Rather, it's the loss of comfortable sharing. Each miscommunication drives the wedge of separation in a little deeper. It widens the gap between who Julian is now and who he used to be. . . . They also bring up fear of the future: a time when Julian may not be able to talk at all, may be unresponsive and mute." Julian expressed this fear openly, telling Ann that he is beginning to imagine how it feels to be someone who can't talk. " 'But my soul is still me,' he said. 'I'm still myself. In a few years, though, I may not be able to speak.[12]

A consolation as they grappled with this fear was Julian's long-standing interest in meditation. "Meditation appealed to him long ago, then as a way of quieting his highly verbal, chattering mind. Now it offers a way of being silent without fear. 'Maybe I'll be a hermit,' he joked with me recently. 'I'll grow my beard and go live in a cave.' 'Not a bad idea,' I'd said. 'I'll live in the valley and bring you food.' " The joke is clearly an expression of their fear of the loss of social interaction that will occur with aphasia and what their relationship will become as he becomes more dependent. The seriousness of this theme emerges in Ann's description of a dinner at a friend's house, when someone spoke of meeting a 65-year-old man in India about to go into retreat and take a vow of silence. "He was 'preparing for silence.' I wondered what he was doing to 'prepare' and wished I'd asked. But the phrase stuck with me. Maybe that's what Julian and I were doing now, each in our own ways: preparing for his silence."[13]

"Preparing for silence" remains a theme throughout the rest of the book, which chronicles both the losses (gradual and relatively minor, since the book covers only a year in the course of an illness that often stretches more than a decade) that she and Julian must endure and the occasional triumphs they share. At the end of the book, though Julian has not become a hermit because of his

Alzheimer's, he has perhaps become something of a holy fool: "A winsome, adorable side of Julian emerges as his intellect dwindles. If I truly let go of needing anything from him, an appealing beauty shines forth. He is as transparent and lovable as a two-year-old, irrepressibly funny, living with fewer and fewer inhibitions. The other day, wearing earphones and listening to Vivaldi, he waved his arms wildly about, 'conducting' the entire *Four Seasons*. Watching him immersed in pleasure brought tears to my eyes."[14]

Lela Knox Shanks, writing of her experience caring for her Alzheimer's-afflicted husband, Hughes, for twelve years at home, also articulated the belief in a self that remains in dementia. "Dementia or disease cannot destroy anyone's essential humanness," she writes; "it is intrinsic to our being, to the life that we are. Dementia is just one of the legions of manifestations of what it means to be human. . . . Being human includes both sickness and wellness, good and evil, old and new, positive and negative, saint and sinner."[15] Even more clearly than Strecker and Davidson, Shanks expresses the idea that dementia does not destroy the true self but in fact helps to reveal it. "Once he had come through the turbulent times of the second stage of AD [when patients' deterioration is severe enough that they lose autonomy but not severe enough that they are unaware of this loss, often resulting in resentment and hostility toward caregivers], a calm seemed to come over him, and the distinctive qualities of his personality resurfaced: his personal charm, his willingness to share, his modesty, his concern for another's safety, and his display of affection." These qualities of Hughes are in fact more apparent and authentic in dementia than previously because dementia has made Hughes more natural. "Hughes, like other demented patients, is much closer to his feelings and basic instincts than so-called normal individuals," she writes. "In the process of becoming civilized and formally educated we often become separated from our feelings and skeptical of our instincts. We put our trust instead in written and visual information, purportedly proven by someone else. But when AD strikes, it is back to the basics. That's all the AD patient has left."[16]

From this perspective, Shanks interpreted many of the behaviors that were typically seen as inexplicable, irrational, and degrading—evidence that the self of the person with dementia is gone or is fast going—as evidence that Hughes was still there. Even the most disturbing of her husband's behaviors appeared to her to be understandable and even rational responses to his situation. Wandering is a reaction to the feeling of being lost and having lost things which the person with Alzheimer's experiences. "Anyone who is lost instinctively tries to find the way back home." Hiding possessions also seemed logical and human to her. "We so-called normal individuals often assume that the person with whom we live has mislaid our possessions when we can't find them. How many of us have called out to a spouse, 'Honey, did you pick up my glasses' or 'Have you put my socks somewhere?' Since AD patients can no longer recognize objects, they cannot 'find' them. When the careprovider 'finds' them, patients then hide their possessions to make sure that no one can take them away again." Spontaneous weeping was a logical response to the immensity of the patient's loss, providing "a safe relief valve for patients at a time when they still have some fleeting understanding that a dreadful disease has seized them and is beyond their control." When Hughes played with children's toys, it was not simply evidence of his deterioration but an example of the persistence of his selfhood. "He has a doll and a bunny he talks to, kisses, and hugs—much as he caressed our four babies. He places them on the couch next to him. When I try to sit on the couch, he frowns, protests, and rushes to scoop the doll and bunny up out of harm's way." On walks, he would continue to react to his environment in a meaningful way. "If there is an obstruction in our path, he presses my hand to steer me away from what he perceives as danger." If they passed a small child, "he would become very agitated, slowing down and looking back, pointing to the child. I am sure he thought the child was lost. Each time I had to reassure him that everything was all right before he would continue on his walk." When she scolded him, he tried to make up with her by taking her hand and kissing it. "His greatest show of affection is when I have settled him in for his toileting. He kisses my hand and cheek. It is as if he knows this is the most difficult part of his care, and this is his way of saying 'thank you.' Clearly, a loss of cognition is not a loss of one's feeling or one's humanity."[17]

Shanks drew on more than religious faith to find the strength to see the humanity in a husband who could no longer remember even his own name. She drew on an African American heritage that taught her to deal with "circumstances in our lives that for a time, at least, we cannot change" and the example and experience of her parents' struggle with adversity, especially her mother, who challenged the constraints of gender as well as race.[18] Both Shanks' and her husband's involvement in the civil rights movement of the 1960s helped her to see how important it was to recognize Hughes' humanity. Like Strecker, Shanks condemns society for failing to see the humanness of the person with Alzheimer's, but her heritage and experience in the civil rights movement allow her to link her critique of this failure to broader struggles for justice and human dignity. Lamenting the widespread tendency to deny the personhood of people with dementia, she argued that "one reason slavery endured for so long in America was that some proponents promoted the idea that the Africans had no souls. It is a dangerous and degenerate society that engages in the dehumanization of any segment of society."[19] Despite her husband's grave dementia, Shanks believed that he continued to be engaged in the struggle that had been such a large and important part of his life, "standing for the human dignity of any individual or group of individuals being treated as less than human." He had been active in the civil rights movement and, as a civil servant and leader of his government union, willing to speak out for the rights of women, religious minorities, and gays and lesbians long before these were popular causes. She saw her husband as still contributing to this larger struggle. "His task at this time is to be an AD patient. . . . He is fulfilling his mission by prompting others to examine current attitudes and policies on what is possible in the care and management of Alzheimer's patients. It is through the profundity of his humanness that he is training me. I believe that he knows somewhere inside that he is still fighting for the human dignity of the individual."[20]

Another source of inspiration for Shanks was social constructionist theory. (Describing herself as an independent scholar and lecturer, Shanks was remarkable among the authors of caregiver narratives for citing a broad range of scholarly, literary, theological, and biblical sources. Most authors of these works cite only the standard popular resources created by professionals for lay caregivers such as *The 36-Hour Day* and *The Loss of Self*.)[21] One of the scholarly sources cited at length is a study by Syracuse University sociologists Robert Bogdan and Steven Taylor of caregivers of profoundly retarded people; these caregivers asserted that the people they cared for were fully human individuals with their own personality, their own feelings and motives, and the ability to respond to love—even though their cognitive abilities were extremely limited and in many cases they were completely unresponsive to communication. Shanks quoted approvingly their assertion that "the definition of a person is to be found in the relationship between the definer and the defined, not determined either by personal characteristics or the abstract meanings attached to the group of which the person is a part." Although the patients in the study sometimes drooled, soiled themselves, and did not talk or walk, their careproviders "viewed them as 'someone like me,' that is, as having the essential qualities to be defined as a human being."[22] Shanks found in the social constructionist position a theoretical justification for her commitment to caring for her husband with a sense of solidarity and respect rather than pity and disgust (though she does not deal with some of the more challenging implications of this, which are discussed in the next section). It gave her a compelling language for denying that there is an essential difference between normal people and those with dementia. No matter how bizarre and disturbing "so-called normal" people sometimes find the behavior of people with dementia, there was no good reason to deny them full moral status as human beings. This commitment, as Shanks recognized, has radical implications for the way our society organizes and distributes care for people with Alzheimer's and all people living with cognitive disabilities.

Still, as with Strecker and Davidson, faith and spirituality were at the center of her narrative. Shanks's faith was not clearly identified as Christian, though she used the Bible with freedom and facility. Hers was a personal faith that evolved under the guidance of two older white women, sisters, who befriended her and her family during the civil rights movement and served as spiritual guides in what Shanks characterized as a "process of self-confrontation and inner development." Prayer was a daily part of her life, and she concluded her chapter on the "humanity of the Alzheimer's patient" by describing the insight she explored with social constructionist theory as first coming as a moment of epiphany through prayer. "Once when Hughes was confused and would not cooperate and I was so angry, I prayed 'God, let me see You in Hughes even when he seems so ugly and unlovable.' The answer came back swift and clear: 'When you see Me in you—then you will see Me in Hughes.' It is in the engagement of other human beings that we are able to experience their humanness and fulfillment of our own."[23]

The importance of Christian religious tradition in these caregiver narratives was paralleled in the work of several philosophers and bioethicists, who, in the realms of the academy and in ethical debates, articulated the claim that people with dementia must be recognized as fully human. David Keck, a church historian and theologian, approaches Alzheimer's as a "theological disease," a disease that raises not only technical problems for medicine and practical challenges for caregivers but profound moral challenges that test the integrity of the church as a body of belief and a community of believers. Keck finds modern theological agendas designed around self-fulfillment and grounded in personal experience inadequate when dealing with the apparent disintegration of the self. It is not only the person with Alzheimer's disease who has "forgotten whose we are" but, too often, the church as well. Alzheimer's requires the church and its believers to remember the central theological concepts of soul and resurrection, to interpolate the suffering of Alzheimer's with Old Testament suffering and the redeeming suffering of Christ on the cross—yet to do so remembering God's love and the promise of salvation. Although Keck sees Alzheimer's as "deconstruction incarnate," like Strecker he sees in the Christian understanding of the soul and God's love grounds for finding the humanity of the patient and for rejecting the notion of an essential difference between the normal and the afflicted. Reflecting on the prayers in another book of spiritual help for caregivers, prayers that often ask God to help with patience, guilt, obsessive fear, and doubt, Keck observes that "we begin to see ourselves as people who resemble Alzheimer's patients. We recognize that our words and actions often make little sense to God. The inadequacy of our words becomes plain, and we see our stammering—and perhaps our suffering in a new light. . . . Through prayer, not only may we begin to see ourselves as Alzheimer's patients, we may also begin to see God in these people."[24]

Bioethicists David Smith and Stephen Post also articulated the challenge posed by Judeo-Christian religious traditions to the distinction between the normal and the cognitively impaired. In bioethics, this distinction is reflected in the model of personhood, which reserves the privileges, obligations, and protections of personhood for those people who are "self-conscious, rational, and in possession of a minimal moral sense."[25] Smith argues that from the perspective of Christian religious tradition, "the appropriate response to suffering is trust in God and solidarity with other sufferers. The radical dichotomy is between God and humankind, not between the demented and the responsible. . . . I can identify with you and myself in dementia precisely because my identity—and yours—lies in a relation to God rather than in the perfection of moral subjectivity. And I can have patience because trust in God makes sense."[26]

Similarly, Post challenges the tendency in what he characterizes as our "hypercognitive culture"—a culture in which the values of rationality, self-control, and mental and physical self-development are emphasized to the exclusion of all others—to think that people with dementia lack any moral significance. "We divide 'them' from 'us,' drawing a line between the rational and the less rational, the unforgetful and the most forgetful, thereby exposing people with dementia to a vulnerability manifest in disregard of their remaining capacities, subjectivity, and well-being. Abuse and neglect of people with dementia is a perennial tendency." Post counters this tendency by an appeal to Judeo-Christian religious traditions that, at their best, "provide a framework for radical equality." He asserts that "Judaism is a bastion of unconditional attentiveness to the frail and needy individual" and quotes from Psalms: "Do not cast me off in my old age, when my strength fails me and my hairs are gray, forsake me not O God." Similarly, Christian ethics makes caring for the weak a special vocation. There is a fundamental equality between persons—all are equally the children of God:

> The chief virtue of this Judeo-Christian framework is that it asserts the moral status and dignity of people such as those with dementia. Their worth as human beings is assessed not in relation to social value, productivity, and rationality, but in relation to the deity, and is therefore absolute.

It is this sense of radical equality, coupled with the ideal of love (solicitous service) for even the most alien neighbor, that can embolden us to cross the boundary line of hypercognitive values in order that people with dementia not be left alone. We should not condone the existential and cultural flight from dementia that is fueled by our only partial acceptance of human realities and our own human selves.[27]

While Smith and Post make a compelling case that Judeo-Christian religious beliefs provide a framework in which the selfhood of people with dementia can be embraced, the caregiver narratives described in this section also reproduced in subtle ways the emphasis on cognitive ability and coherence as the essence of selfhood. The claims for the continued selfhood of the person with dementia in these narratives rested on a romantic notion of the self in which some essential element of a person's identity remained intact no matter how much brain deterioration he or she suffered. The actions of the person with dementia—who is possessed of an intact, if challenged, self—could communicate that selfhood to those who were prepared to receive it. Thus, in these accounts the selfhood of the person with dementia still hinges on the exercise of cognitive abilities, albeit at a level that much of society would not recognize as meaningful. For example, when Theresa Strecker realizes that her father "may have forgotten our names, but not that he loved us," the key point in this story is that at some level, deeper than words, he continues to recognize and love them.

"HENRY WAS A LAWYER": THE RECONSTRUCTED SELF IN SOCIAL CONSTRUCTIONIST THEORY

Religious belief and spirituality have not been the only bases for asserting the humanness of people with dementia. Lela Knox Shanks found that social constructionist position could help her sustain her commitment to regard her husband as an intact person struggling with a chronic neurodegenerative illness rather than as a person who had already been "lost" to the disease, a person who was in fact no longer really there. Shanks' reading glosses over the more radical implications of social constructionist theory. Bogdan and Taylor (the authors she cites) were not creating a special account of selfhood to describe the situation of the profoundly retarded and their caregivers. Rather, they were applying "toward people who have often been denied their humanity by being defined as nonpersons" a perspective on the social creation of selfhood in general which had been developing in American sociology since the work of Charles Horton Cooley and George Herbert Mead in the first two decades of the twentieth century.[28]

According to this perspective, humanness is not simply an intrinsic quality possessed by people at birth or any other time. Social interactions were a key element of selfhood as well. Selfhood in the social constructionist account was largely an interpretation of human behavior, an interpretation that was instantiated through social interactions. Bogdan and Taylor concluded their article by pointing out that it would be easy to dismiss their conclusions by claiming that caregivers are deceiving themselves when they attribute human qualities to the profoundly retarded people they cared for, but the authors argued that it is just as possible that those who dehumanize people with severe disabilities are deceiving themselves. "After all, no one can ever prove . . . that the assumption of common experience is anything but an illusion. What and who others, as well as we, are depends upon our relationships with them and what we choose to make of us."[29] Selfhood, then, whether we are talking about "normal" people, retarded children, or people with Alzheimer's, hinges on no more nor less than the interpretation of behavior—a concept somewhat at odds with the idea articulated by caregivers in the previous section that people with dementia continue to possess the essential humanness that God bestowed on all people.[30]

Nonetheless, social constructionist accounts of Alzheimer's disease were indeed overwhelmingly committed to laying the theoretical basis for reconstructing the selfhood of the person with Alzheimer's disease—in the words of Bogdan and Taylor, to creating a "sociology of acceptance" that emphasized the ability of the well to sustain the humanness of the ill and disabled, as opposed to a "sociology of exclusion" that viewed stigma and labeling of the deviant as an inevitable part of the illness experience.[31] By emphasizing the constructedness of selfhood, these critics of the notion that selfhood was lost in Alzheimer's blurred the boundary between "normal" and "demented," between self and nonself. A person, whether normal or demented, possessed an intact self only to the degree that others interpreted his or her behavior as indicative of selfhood. But what were the criteria for such an interpretation? These, of course, were themselves open to debate, and a debate that involved fundamental political and moral issues. Constructionist critics of the representation of people with Alzheimer's in terms such as "the living dead" called for a revised set of criteria that would define people with dementia as human. But in so doing, they revealed the problematic nature of selfhood in general.

Anthropologist Elizabeth Herskovits has argued that what is at stake in the burgeoning literature on the self in Alzheimer's disease "is our very notion of what comprises the self and what constitutes subjective experience." We can see this more clearly by examining in some detail two differing accounts of the social construction of self in Alzheimer's disease. One account is essentially a theorization of the prevailing view that selfhood is lost in dementia because of disease processes, whereas the other claims that the loss of self is not a product of the disease per se but of the invalidation and stigma the person with dementia endures. For reasons to be discussed, the former view is rarely expressed among social constructionists,[32] while some version of the latter has become practically conventional wisdom among many who study social aspects of Alzheimer's disease.[33] Both accounts use very similar sorts of evidence and methods—ethnographic study of the interaction of people with dementia and caregivers in adult day care centers and nursing homes. Both of these interpretations also rest on essentially the same body of theory—symbolic interactionism or ethnomethodology, or both—though they make different assumptions about where to draw the line between social processes and human subjectivity.

The crucial element of the symbolic interactionist/ethnomethodological theorization of the self is the distinction between an inner, personal self and a publicly presented self or series of selves. This line of theory is usually traced back to George Herbert Mead's distinction between "the 'I' which is aware of the social 'Me. . . . The 'I' is the response of the organism to the attitudes of the others; the 'me' is the organized set of attitudes of others which one himself assumes. The attitude of the others constitute the organized 'me,' and then one reacts toward that as an 'I.' "[34] Perhaps the best-known articulation of this distinction, or at least the version most influential among the authors considered here, was the extended metaphor employed by Erving Goffman of selfhood as a drama, requiring the participation of a performer and an audience.[35] Goffman distinguishes between the self as *performer*, "a harried fabricator of impressions involved in the all-too-human task of staging a performance," and the self as *character*, "a figure, typically a fine one, whose spirit, strength, and other sterling qualities the performance was designed to evoke." The performer is obviously the personal self, or Mead's "I," and character the public self, or Mead's "me."

Goffman's account focuses more attention on the "me," or public self, as the product of social processes, and its status is clear: "A correctly staged and performed scene leads the audience to impute a self to a performed character, but this imputation—this self—is a *product* of a scene that comes off, and is not a *cause* of it. The self, then, as a performed character, is not an organic thing that has a specific location, whose fundamental fate is to be born, to mature, and to die; it is a dramatic effect arising diffusely from a scene that is presented, and the characteristic issue, the crucial concern, is whether it will be credited or discredited."[36] But Goffman is much less clear about the nature of the personal self, though he stresses that as performer the personal self has agency and self-awareness. "He has a capacity to learn, this being exercised in the task of training for a part. He is given to having fantasies and dreams, some that pleasurably unfold a triumphant performance, others full of anxiety and dread that nervously deal with vital discreditings in a public front region." But in Goffman's account it is not clear whether this personal self is intrinsic and essential to a person (whether it is fated "to be born, to mature and to die," as the social self is not) or whether it is contingent in some way on the success or failure of performances. Goffman is simply ambiguous: "These attributes of the individual *qua* performer are not merely a depicted effect of particular performances; they are psychobiological in nature, and yet they seem to arise out of intimate interaction with the contingencies of staging performances."[37] This ambiguity is the focal point of the two differing interpretations of selfhood in dementia. Each position is based on a different a priori concept of the personal self, and as a result the two positions reach different conclusions about whether the "performances" of the person with dementia can sensibly be credited.

Publishing in the same year as the study by Bogdan and Taylor, Andrea Fontana and Ronald W. Smith from the University of Nevada at Las Vegas shared Goffman's emphasis on the agency and self-awareness of the personal self. To be functional, the personal self had to demonstrate a mastery of the various roles and rituals it needed to perform to establish a credible public self or selves. Alzheimer's demonstrably disrupted the personal self's mastery of these roles and rituals, although the illusion of selfhood might be maintained as habitual behaviors were repeated in ways that were interpreted by caregivers as credible performances of self. In people with Alzheimer's, "the self has slowly unraveled and 'unbecome' a self, but the caregivers take the role of the other and assume that there is a person behind the largely unwitting presentation of self of its victims, albeit in reality there is less and less, until where once there was a unique individual there is but emptiness." Citing Goffman and Harold Garfinkle, Fontana and Smith explained that "everyday reality rests on the routines (the ethnomethods) by which the members of society accept each other. We, *qua* members of society, accept the 'normal,' often taken-for-granted routine ways in which others 'pass' and, where it is necessary, we 'fill in the gaps.'" But this "filling in the gaps" or "normalizing of competence" was highly problematic in the case of people with Alzheimer's disease "because they retain proper social 'forms,' but the 'content' of their actions becomes increasingly meaningless."[38]

Fontana and Smith's account was remarkably exclusionary. Using much the same sort of language as popular representations, their article substantiated the image of the Alzheimer's victim as a hollow shell. They found no grounds for interpreting the occasionally appropriate behaviors of Alzheimer's disease victims as evidence of an essentially intact, if significantly challenged, selfhood, as caregivers sometimes (and other social constructionists often) did. These behaviors were merely vestigial traces of deeply ingrained social practices, not evidence of agency or individuality on the part of the person with Alzheimer's. Occasionally, these vestigial traces would click in when a habitual situation presented itself, and the person with dementia might appear normal for a moment or even a brief period. Caregivers often interpreted these as evidence of a functioning self. But as soon as the situation becomes novel, calling for a creative and masterful manipulation of these social forms, the emptiness of the Alzheimer's victim becomes apparent. In the early stages of the disease people with Alzheimer's "continue to interact on the surface as if they were sentient beings, while they are in fact losing the rational 'content' of their self and relying increasingly on 'forms' of sociability, deeply embedded in them through socialization experiences. . . . To the extent that they manage to accomplish these routine tasks they are treated as a full-fledged self." This facade of normalcy "proceeds until it becomes known that the form is not accompanied by content on the part of the Alzheimer's victim. Then the interaction can become a problematic matter for the non-Alzheimer afflicted participant. Still, some semblance of normal interaction can continue because of the rituals used by the actors and the normalization processes employed by the non-Alzheimer person for the victim." For example, the authors describe how "routine actions are important in allowing the victims to be seen as competent, albeit often closer inspection will reveal the emptiness behind the façade." Thus a patient sitting at the desk when the phone rang picked it up and answered "hello," giving the appearance of competency, but "then held the phone out in front of himself, looked at it quizzically, and said confusedly: 'What on earth is this?' and handed the strange object to the nurse." Similarly, normal talk creates a sense of competency that careful scrutiny often reveals is hollow, as when the authors "asked Don, a patient, 'How was the visit to the doctor yesterday?' thus reminding him that there had been such a visit. He replied: "He couldn't tell me anything. Said he had to talk to some of the others about my condition. Makes me feel like I'm pregnant.' The appointment was actually for his wife."[39]

Fontana and Smith regarded even complex social behaviors accomplished by people with dementia as "empty" in this sense. One member of the research team observed a card game among patients and volunteers. Sarah was "one of the more animated players," and a volunteer explained to the researcher how remarkable it was that she was playing at all because her dementia had progressed quite far, to the point where she wandered from the unit and got completely lost trying to find her way "home," actually, a home that she had not lived in for many years. Nonetheless, "Sarah still plays spades with her patient friends, although even they seem aware of her waning skills. Her good friend Joan always sits to the left of Sarah to grab an occasional 'good card' that Sarah mistakenly discards."[40] By Fontana and Smith's account, Sarah's marginal ability to participate in a card game, an activity that is endorsed as evidence of retained selfhood by others, is simply a residual effect of socialization, not an indication that Sarah is "really there," that she is a functional self.[41]

Fontana and Smith listed several possible reasons that caregivers might want to "normalize" the increasingly bizarre behavior of people with dementia. Family members may be "merely trying to normalize the new and odd behaviors that they do not understand," or they may simply be deceived by "the apparently appropriate social customs being displayed by the victim." Denial is another possible explanation. "Out of devotion or love the caregiver may have difficulty admitting to self and others that the individual has a deteriorated mental condition." Overlooking or ignoring incompetence might also, the authors thought, be a way of managing the practical issues of caregiving. Related to this, caregivers might normalize the conduct of people with Alzheimer's in order to avoid frightening others or to avoid embarrassment. Whatever the reasons, Fontana and Smith conclude that "there is little doubt that the self of the Alzheimer's disease patient is interpreted and mediated by friends, relatives, and staff members. Often these others protect and speak for the victims and use many devices to defuse their misconduct. They assume or pretend that the individual is 'competent' and in so doing preserve the social self of the patient."[42]

Writing a few years later, Georgetown University psychologist Steven Sabat and Oxford University philosopher Rom Harré began with a different notion of the personal self, and this led them to much different conclusions.[43] In their account, the normal personal self is "the continuity of one's point of view in the world of objects in space and time" and is "coupled with one's sense of personal agency, in that one takes oneself as acting from the same point." Personal self does not lie in the mastery of particular social forms but rather in the simple existence of a coherent point of view. Where one's social self or repertoire of selves must conform in content to the person—types recognized by the community in order to be realized—the personal self is merely a formal unity, "a point in 'psychological space.'" Personal self is "a structural or organizational feature of one's mentality," itself devoid of content. "However the content of our thoughts and feelings may change, we are intact as persons if these are organized into one coherent whole." As such, the personal self does not depend on the cooperation of other people for its existence or on an intact memory or a high level of cognitive function. The existence of a personal self is confirmed by the coherent use of "first person indexicals," that is, the use of pronouns (I, me, mine, myself) to make statements that could be used in performance of a social self. Coherent use of first-person indexicals does not even require that the speaker remember his or her own name, so long as the speech frames his or her discourse (no matter how confused) as coming from a personal identity. From this perspective, Sabat and Harré were puzzled by the notion that the self is lost in Alzheimer's disease. "If the A.D. sufferer can be shown to employ first person indexicals coherently in his or her discourse, on the constructionist account, the A.D. sufferer has displayed an intact self."[44] In their analysis of observations and interviews gathered at an adult day care center, the authors showed that even people who had severe dementia continued to use first-person indexicals. Even the loss of the ability to speak at all did not obliterate the personal self, for people in this condition still confirmed its existence by using indexical gestures not unlike the formal indexicals of American Sign Language.[45] The authors considered the possibility that the appropriate use of pronouns could simply be a habit, as in Fontana and Smith's account (which Sabat and Harré do not cite), but they argue that the competence displayed is on too high an order for this to be the case.

As an example, they describe a series of interviews with a patient, J.B., a man with Alzheimer's who spent his work life as an academic. In the early interviews, J.B. employed first-person indexicals fluidly and was able to engage relatively well in conversational give and take:

> S.R.S.: You don't think that you've lost your sense of self?
>
> J.B.: In what respect?
>
> S.R.S.: You know who you are, you know that you are.
>
> J.B.: Yeah, definitely. I know who I am. And sometimes I have to fake, um, as to people that I deal with back and forth.
>
> S.R.S.: What do you have to fake?
>
> J.B.: Uh, I have to fake for, uh, course I feel I could, could have done more. Can I do better now? I don't know.

Sabat and Harré argue that J.B.'s use of first-person indexicals in this passage is too complex and fluid to be merely a habit. Although the content of these statements is somewhat unclear (it is not entirely clear, at least from the extract the authors provide, what he has to fake and why), J.B. not only uses the indexical "I" appropriately but does so in a complex construction in which it is embedded in another indexed statement that refers to self-presentation (the "I's" need to fake for others). In their analysis, J.B.'s speech indicates "the high level of his control of some aspects of the 'mechanisms' of self-presentation." Although J.B.'s use of indexicals is "empty" in the sense of being purely formal, he employs them with an understanding of the way they are related to the presentation of a social self. This is clearer when, in response to the researcher's question about what he thinks about his Alzheimer's, J.B. replies that "I'm mad as hell . . . constantly on my mind . . . which may or may not screw up your project." In this statement, Sabat and Harré see J.B.'s formal indexical completed by the content of the sentence, a reference to his possible social self as a "scientific collaborator" on their project.[46]

In later interviews, J.B.'s condition had deteriorated significantly. His emotional turmoil had deepened; he was unable to carry on a conversation with as much give-and-take; he could hardly walk, dress himself, or find his way around without assistance; and he appeared to be in great pain. Yet his ability to employ indexicals remained unimpaired (in the remaining quotations from J.B.'s interviews, I have combined J.B.'s statements and omitted the words of S.R.S.): "I'm so pissed off. . . . Let's take (the social worker). Could you and I, just the two of us, get, get, on. on someplace to the idea of what you and I do . . . ? She says I don't think you should be working anymore. And then I don't think you should be doing this." Sabat and Harré comment that J.B. "displays the same discursive ability with indexicals as the passages from the earlier interviews" and a continued understanding of the ways in which these indexicals are related to the performance of a specific social self, that of the "scientific collaborator," which the social worker evidently would not legitimate.

But if the personal self remained intact even into the advanced stages of dementia, the social self or selves were highly vulnerable to stigma and invalidation. "One's display of the characteristics of a certain persona enters 'social space' only in so far as it is recognized, responded to and confirmed in the actions of others." Thus the loss of self in Alzheimer's disease was, according to Sabat and Harré, the loss of a social self. It was not primarily caused by the disease but by "nothing more than the ways in which others view and treat the A.D. sufferer." As an example, they cited the way an individual with Alzheimer's was introduced by a member of the staff to another person at the day care center: " 'This is Henry. Henry was a lawyer.' Henry interrupted the proceedings by saying, 'I *am* a lawyer.' " Although Henry was no longer practicing law, he "had not been disbarred, nor had his degree or years of practice and achievements been nullified by his developing Alzheimer's." In this brief exchange, Henry struggles not only with the difficulties caused by the disease but also with the difficulties brought about by the refusal of others to recognize an aspect of his self that had been crucial to his identity for most of his life and which he still needed—perhaps even more—as he developed dementia.[47] To paraphrase Chris Gilleard, another social constructionist whose position was very similar to that of Sabat and Harré, Alzheimer's disease involved as much losing your place as losing your mind.[48]

As an alternative to delegitimating aspects of the selfhood of the person with Alzheimer's, Sabat and Harré describe the denouement of J.B.'s case. In his final interview, J.B. made it clear that he wanted his status as a "scientific collaborator" on the authors' project recognized:

> I have the feeling, some feeling that I don't necessarily have status, um, because its not really something that I'm piddling with. And you know I feel a way is that, I feel that this is a real good, big project and I'm sure you do too. This project is a sort of scientific thing. . . . Because I've has this, people talk to me and then I'm blum, blum, blum. What do I do? Well, blum, blum, blum. So it may not mean very much. Maybe it's nothing. Because others go with stature and what I feel, I think my God this is real stature to do! This is maybe picky, picky stuff. . . . That is stature and who can be attached to somebody, uh, in an agreeable thing. Somewhere, where we can do status. If somebody calls me and says what are you doing and then I write this thing, what is it? Now it could be a long, long thing in the project from your university or something else. Do you think I'm silly?

Once the authors understood that J.B. wished to link this "last spurt of academic effort with the final stage of his moral career," they arranged to have a letter from the dean of the College of Arts and Sciences commending J.B. for giving of himself unstintingly to help the investigation of the abilities that remain intact despite Alzheimer's, and the letter became a source of great pride to him. Thus the "academic self" that J.B. had performed throughout his life was "jointly constructed once again. By virtue of the social force of the letter of commendation J.B. was positioned, not as a helpless and confused A.D. sufferer, but as one who had a contribution to make to science even in the throes of A.D."[49]

Sabat and Harré's account of selfhood in Alzheimer's disease is precisely the opposite of Fontana and Smith's. The latter presumed, on the basis of the inability of people with Alzheimer's to master the content of social interactions, that the disease causes a deterioration and eventual loss of a personal self which is ameliorated to some degree by caregivers who continue to grant the demented the social status of selfhood on the basis of their (increasingly shallow and meaningless) performance of social rituals. Sabat and Harré presumed, on the basis of the continued ability of even people with severe dementia to appropriately employ formal first-person indexicals in their discourse no matter how confused its content, that personal selfhood remains intact in the disease but that the marginalization of people with dementia by others results in a loss of the social aspects of selfhood. Both pairs of authors agreed that caregivers sometimes act to legitimate and thus maintain the social aspects of selfhood of people with dementia, but they disagreed about the significance of this. Fontana and Smith found it surprising how often caregivers act this way but viewed it as irrelevant to the loss of personal self which occurred with the disease. Sabat and Harré lamented that such supportive interactions do not occur more often, for in their account the personal self is not destroyed by the disease; rather, social aspects of selfhood that are crucial to a coherent, stable identity are destroyed when others act as though the person with dementia is an "empty shell."

It is obvious that these differing accounts cannot be resolved by an appeal to evidence, for either of them could easily reinterpret the evidence supplied by the other as support for its position.[50] Rather, as Herskovits points out, these differing accounts of selfhood turn on commitments to a certain model of the self. But there is more at stake here than theory. Herskovits argues that these debates are driven by a broad social concern about the prospect of aging. Paraphrasing Geertz, Herskovits suggests that "stories of Alzheimer's are stories we tell ourselves about being and becoming old."[51] While I think Herskovits is right, I would go further to suggest that the concern over the issue of subjectivity in dementia goes beyond anxiety about aging; it also reflects concern about the problematic nature of selfhood in general. Theologian and church historian David Keck has called Alzheimer's disease "deconstruction incarnate" because it "subverts our narratives . . . and challenges and relativizes all of our assumptions about language, meaning, and humanity itself. A true first-person, end-stage Alzheimer's narrative—if it were physically possible for someone who cannot speak, type, or hold a pencil—would be the ideal deconstructionist text."[52] But "deconstructionist theory" does not apply only to dementia, as Keck seems to want it, but to our entire culture. If the critical theories Keck invokes are taken seriously, we must all confront the subversion of narrative, the instability of language and meaning, and a fractured, contingent selfhood. The stories we tell about Alzheimer's are ultimately the stories we tell about ourselves in a culture characterized by the subversion of narrative, the contingency and instability of language and meaning, and an often fractured, disjointed experience of subjectivity.

Thus, Fontana and Smith can be characterized as committed to a modern self and Sabat and Harré to what, for lack of a better term, can be called a postmodern self. Fontana and Smith are committed to a model of self that prizes cognitive mastery of the social world: a person is a self because she knows, in very concrete terms, who, what, and where she is in the world. It is the prevalent view of our society. Sabat and Harré are committed to a model of self rooted in the inter-subjectivity of the social: a person is a self when he is connected to a world that will accept him no matter how grave his failings, no matter how confused and forgetful he may be about the particulars of who, what and where he is.

"LIVING HALFWAY":
BETWEEN MODERN AND POSTMODERN SELVES

The conflict between these two models can be seen in concrete terms in the published memoirs of people diagnosed with Alzheimer's disease.[53] The authors of these books are not attempting to articulate a theory of selfhood, and they are hardly inclined to see themselves as champions of a "postmodern self." All are "exemplary victims" in the sense described in the previous chapter—solidly middle class, responsible, productive, and successful before the disease disrupted their lives—and the stories they tell are of their struggle to maintain this identity as the disease progresses. But in trying to make their experience of Alzheimer's disease somehow meaningful, they are compelled to accept the loss of qualities and abilities that are at the core of the notion of modern self, to find value in aspects of themselves that they had previously ignored, and to come to terms with a radically uncertain future in which not only their fate was contingent but their ability to understand that fate as well. Although they are far from explicitly embracing the notion of a "postmodern self," their experience of dementia challenges them to move beyond the confines of the conventional idea of selfhood in our society.

If one is looking for the "ideal deconstructionist text" that David Keck imagined an Alzheimer's narrative would be, these books are a disappointment. All have been written by people with early-onset Alzheimer's (all being diagnosed before the age of 60). All but one of the books were written while their authors were in the early stage of the disease, and these books take the form of conventional autobiographies—so conventional, in fact, that one could perhaps be excused for questioning whether they are "authentic" dementia narratives, that is, whether the authors actually do have Alzheimer's disease and, if so, whether they received extensive editorial help to produce so coherent and focused a narrative.[54] But the value of these books is not that they give us some sort of voyeuristic window on the authentic, unvarnished Alzheimer's experience but that, in trying to come to terms with the Alzheimer's experience, they trouble the boundary between dementia and normality, forcing us to rethink the question of what it means to be a person. One author—Cary Smith Henderson—was in the middle stages of dementia and wrote by dictating his thoughts into a tape recorder. The tapes were later edited by his family and the photographer who illustrated the book. As a result, the book appears as an unplotted collection of free-floating epigrams—what Henderson himself calls the "anecdotal career of an Alzheimer's patient."[55] Nonetheless, even though Henderson's overall narrative is chaotic, his musings contain the same consistent and coherent theme that is more formally expressed in the more conventional narrative structure of the other books.

The titles of three of the conventional autobiographies—*My Journey into Alzheimer's Disease, Living in the Labyrinth,* and *Show Me the Way to Go Home*—indicate that the story of the authors' illness is told in the shape of a journey. As described by Arthur W. Frank, the quest narrative is the story of a long and difficult journey that ends with the hero receiving a boon. When an illness story takes the form of the quest narrative, the experience of illness is seen as a journey that is resolved when the narrator receives the boon of insight. Communicating that insight restores a sense of meaning and purpose for the author and reconnects him or her to the world.[56] All the books follow this basic structure, beginning when the diagnosis of Alzheimer's disease ruptures the expected flow of the authors' lives, followed by a chaotic period of denial, evasion, and isolation and finally a resolution in which the authors come to accept their situation, restore vital connections to those closest to them, and gain a renewed sense of life's potential through the very act of creating their narrative.

In *My Journey into Alzheimer's Disease,* Robert Davis, pastor of one of Miami's largest churches, frames the story of his illness as the culmination of a life of faithful service to God. He recounts the story of his life in some detail, his childhood of poverty, his call to the ministry and years of successful service. Yet when Alzheimer's forced him to retire from the ministry at the pinnacle of his career, he felt plunged into darkness, unable to feel God's loving presence in his life. The resolution of his narrative is a profound and sudden experience of his dependence on God's grace. "Now, instead of my reaching out to Christ by prayer, intellectual determination, sheer bull-headed faith, or by aggressively claiming the promises of Scripture, Christ reached down and held me close to him," he wrote. "From now on, my lot in life would be to be especially held by the Shepherd, letting him fully care for me."[57] Of course, this insight did not relieve him of all the burdens and frustrations of Alzheimer's disease, but it gave him the means of coping with his losses and moved him to try to help others by telling the story of his illness. "I want to be the voice for all those victims who lost their ability to communicate even before anyone knew what was bringing on all these devastating changes," he writes. "Many of these victims were written off by their families. . . . I want to shout, 'Be gentle with your loved ones. Listen to them. Hear their whispered pain. Touch them. Help them stay in touch with God. Let them draw from your strength.' "[58]

In more secular terms, Larry Rose tells a very similar story. He presents himself in *Show Me the Way to Go Home* as a rugged individualist struggling to find a purpose to life as he confronts his increasing dependency. His book begins, as in one way or another all these narratives do, with an account of getting lost on his way to a familiar destination—in this case his mountain cabin, where he vacations alone. After several other alarming lapses, Rose confronts the possibility that there is something badly wrong. "For the first time in my life, doubt stirred somewhere deep inside," he writes. "It was unsettling and unnerving, and for a time, I felt the uncertainty of a person experiencing a hurricane or a tornado for the first time; the terrifying sensation that comes on realizing that what should be firm and solid is no longer so, and cannot be relied upon."[59] A few months later, Rose received the diagnosis of Alzheimer's after reluctantly consulting a doctor about his memory problems and was filled with doubts: "I try to face reality. Will I soon forget who I am? Is there a reason for all of this? Why am I living if there is no purpose to life at all? Will I soon be leading an empty existence? No, I can't be thinking that. A life is never wasted. Even in this helpless state, there has to be a reason. I know that even in the most hopeless situation, there is still a possibility for growth."[60] Rose finds no easy answers to these questions, but at the end of the book he is resolved not to allow the uncertainties to paralyze him. At a plateau in his deterioration, he acknowledges that "I could stay on this plateau for ten years or ten minutes. There is no way of telling. I am going to live every minute like it was my last. If my condition should worsen, no one can say I didn't give life everything I had, that I didn't try everything possible."[61] Rose's boon is to find the strength and courage to continue the struggle to preserve his ideal of selfhood, even as he can no longer evade the fact that it will inevitably slip away from him.

Living in the Labyrinth is Diana Friel McGowin's story of coming to terms with the losses of Alzheimer's disease. Throughout the book, the central drama is her frantic, futile efforts to hide the signs of her disease and pass as normal. "I recoiled in alarm at each additional loss of memory, and concentration. But I would not confide in family members as each additional loss occurred. Instead I continued playing a camouflage game of 'I've Got a Secret,' long after everyone knew. Each loss was magnified in my fear and grief, yet I continued my foolish attempts at deception."[62] The drama is resolved when she learns to become less preoccupied with her losses and more with what she has retained, allowing her to live the remainder of her life as fully as possible. The culmination of this is her leadership in the creation of a support group for people in the early stages of dementia, which spreads across the country. (Rose in fact includes a chapter in his book which describes how he saw McGowin on TV and subsequently contacted her, resulting in an enduring friendship.) "As I read over this book, taking the reader from my first symptoms through diagnosis and denial, I realize how far I have come psychologically, if not intellectually," McGowin concludes, describing how the act of writing this book and an ongoing journal gave her the means of coping. Still, the disease continued to encroach on her life. For example, she describes how one of the secondary symptoms of her disease—her heightened olfactory senses—played havoc with her life, causing her to search for days for the source of a cat urine smell that no one else could detect. On the other hand, such losses had compensations as well: "I can sometimes enjoy the sweet fragrance of night-blooming jasmine, when no one else can. It is my own private and enjoyable sensation. I am aware . . . that this means the disease is, indeed, progressing within me. I can handle it however. Have you ever experienced and enjoyed the fragrance of night-blooming jasmine?"[63]

Jeanne Lee's Alzheimer's narrative is rooted in a sometimes painful past that began when she was sexually abused as a child and included bouts of depression, alcohol and drug abuse, and three failed marriages. Yet through these difficult periods in her life she was able to successfully manage a number of highly responsible and challenging jobs (sometimes more than one at a time) while raising five children. It is when, at the age of 49, she can longer adequately meet demands at work that Lee begins to suspect she has Alzheimer's. Like McGowin, she tells of her struggle to keep her impairments hidden from others and her shame when she no longer can. But Lee is initially given a diagnosis of stress and alcohol-induced dementia, which pulls her deeper into a spiral of guilt and self-punishing behavior. The turning point in her life comes when, with the help of a gifted and caring psychiatrist, she uncovers and works through her childhood sexual abuse and stops her self-punishing behavior such as alcohol abuse. She and the psychiatrist had hoped that her confusion and forgetfulness would improve when her depression lifted. After two years of therapy, the depression was clearly much better, but there was little improvement in her judgment and short-term memory. Further testing led to the diagnosis of probable Alzheimer's disease, which Lee describes as lifting a tremendous burden from her shoulders. "You may find this strange, but I consider my Alzheimer's diagnosis to be the best news I'd heard in my life," she writes. "Now, finally, after years of torturous uncertainty, I had something that I could tell others. What I had was a fairly common disease, that most people have at least heard of, even if most don't fully understand it. I wasn't crazy. I wasn't stupid. I could quit trying so hard to cover up my forgetfulness. I could now write about it and talk about and laugh about it. It was okay if I made mistakes. Now I could say to others, 'I have Alzheimer's, and I'm doing the best that I can do.' "[64]

Lee is able to accept the diagnosis and finds self-worth in the use and enjoyment of the abilities she retains rather than mourning what she has lost. Yet she continues to struggle with a society that still misunderstands and stigmatizes people with dementia. Throughout the book, Lee recounts the way that the stress and frustration she feels when people are impatient with her mistakes compound her confusion. On the other hand, she is also frustrated by people who will not let her be independent or who seek to deny the real problems she faces. "People need to understand to let you have your space and make your mistakes," she explains. "I can live with it quite nicely, as long as people don't expect too much and aren't constantly trying to convince me that there is nothing wrong. Don't try to teach me things I used to know and be good at, and don't overprotect, or act like it's some terrible thing that we should pretend isn't there. . . . I've accepted that I'll never be the way I once was, but I'm still a person. I just want people to understand that it's not easy, but life goes on."[65] Like the other authors of dementia memoirs, Lee finds redemption in narrating her experience in the hope that it will help others who struggle with dementia and, as importantly, help those who do not to better understand. Lee's book is ultimately a plea that people with dementia be given the respect and support they need to sustain their lives, as indicated by its title—*Just Love Me*.

Thomas DeBaggio's memoirs are the most ambitious of these books. They are spread over two volumes—*Losing My Mind* and *When It Grows Dark*—evidently composed in a year or two after he was diagnosed with Alzheimer's disease and completed with help from his wife and son.[66] In each volume DeBaggio interweaves two narratives: his memories from earliest childhood through the present, and his struggles to capture those memories and understand the nature of their destruction as his dementia progressed. The first volume tells the story of his coming of age in Arlington, Virginia, during the 1960s; his youthful literary ambitions; the difficult relationship he had with his parents as he struggled to make his own way in the world as an artist, a writer, and later a journalist for a daily newspaper; and the joys and struggles of his early years of marriage and fatherhood in the early 1970s. As a journalist, DeBaggio witnessed much of the social and cultural turmoil of the period, covering, for example, the bitterness of segregation and racial strife in Wilmington, Delaware. But he grew frustrated with what he saw as the ethical compromises mainstream journalism required, as well as the harassment that he and his wife received for their involvement in civil rights and other causes. After a short stint self-publishing a well-regarded but financially untenable independent newsletter, he left journalism altogether and eventually developed his longtime love of gardening into a successful business growing and selling herbs out of his backyard. The second volume tells the story of the long struggle to develop this business, the joys and insights that gardening brought him, and his deepening relationship with his wife and son.

DeBaggio is proud of what he has accomplished with his business, but he also regrets the life he has to a large extent given up, the life of words and ideas. Although he is initially shocked and horrified by the diagnosis, he eventually sees it as an opportunity. "After those weeks of silent introspection, I realized this slow-moving disease offered a great opportunity to me. The Alzheimer's diagnosis freed me at last to write seriously and well if my gift of language did not fail me."[67] The central drama of these books is DeBaggio's struggle to tell the story of his rich life even as the memories he seeks and the words with which he must capture them slip away. In language and images that seem to reflect the trope of the loss of self in the broader popular discourse on Alzheimer's described in chapter 5, DeBaggio frequently suggests that his failing memory and diminished language destroy his very selfhood. "Am I anything without my memory and the simple skills of reading and writing I learned in childhood?" he wonders. In many passages, he answers this question in the negative: "With failing memory, it is difficult to write long passages without getting lost in words. Where does the story go? Why does the pencil tremble? I see only the structure of words, their meaning elusive. I am often able to write only a sentence of two, enough to sketch what was to be brawny and complex. Do you understand I am not dying, just disappearing before your eyes?"[68] Yet in other passages he sees his struggle as lending a dignity and purpose to his life which give him joy despite these losses. "I have begun to adjust my life so each day has a structure to it, and a purpose; to enjoy every minute I can and to focus on the work I love with herb plants, and with words. I want to write the truest sentences I can in the hope my words give others the sense of struggle and joy I feel."[69] DeBaggio recognizes that as his own memory fails his existence will persist in the memory of others. "Although my memory is crumbling into obscurity, I am a memory for someone, living again in their recollection. . . . Our immortality, such as it may be, is not contained in what we dreamed or the secrets we kept; it is how our loved ones remember us."[70] Alzheimer's provides no easy redemption for DeBaggio; in the final pages of his memoir he describes it as "a deadly hole in my life, a deep cistern in which I tripped unknowingly." As he waits for death, he longs only for a final farewell from his wife and son. But the very last words of the second book trail into elegiac abstraction: "White gulls sweep from the sky. In a moment they soar to new heights on warm, familiar currents, as cars speed beneath them." He describes a solitary man, walking along the sidewalk as he has for decades. Is it DeBaggio? His consciousness seems merged with the landscape, becoming perhaps someone else's memory. "As the walker comes to a turning point, he adjusts his cap low on his head, preparing to look into the bright rising sun. Balloons dance over electric lights high above the street."[71]

Cary Smith Henderson's *Partial View* is a much darker, frightening representation of the Alzheimer's disease experience than these conventional autobiographies. Henderson was evidently unable to sustain a line of thought for more than a few sentences. His tape-recorded ruminations are arranged more or less by topic, but these topics are not identified or marked in any way; nor is there any indication of their temporal relationship to one another. Missing too are any conventional signposts orienting the reader to time or place. The disorientation is heightened by Nancy Andrews' photographs, which frequently use dramatic alterations of lighting and perspective to depict Henderson contending bravely in a world in which everyday objects became strange and threatening. The reader is meant not only to read a description of the chaotic world of Alzheimer's disease but to experience it in the confusing structurelessness of the narrative itself. Deprivation and loss are the main themes of Henderson's narrative fragments. "I've been thinking about myself. Some time back we used to be, I hesitate to say the word, 'human beings.' We worked, we made money, we had kids, and a lot of things we enjoyed. We were part of the economy," he says in one passage, speaking evidently for all people with Alzheimer's. "I was into all sorts of things, things that had to do with charities, or things that had to do with music. Just a lot of things I did when I was, I was about to say—alive—that may be an exaggeration, but I must say this really is, it's living, it's living halfway."[72] Yet despite the fragmentation of the narrative and the emphasis on deprivation and loss, Henderson expresses essentially the same themes that the four conventional autobiographers do. Like the other authors, he struggles to integrate the losses of the disease with his sense of self. "Alzheimer's is a lot of stress, mainly because you know what you have been earlier and you know very well you're not that good now and it's real hard to reconcile. There are times when I've felt like destroying things—sometimes I want to cuss," he notes. "We do need to be taken care of. I've learned pretty much the parameters of my life—how far I can go and what I can expect to do and I can live with—I think. I'm an onlooker to life not, not a participant. An onlooker and a philosopher." Like the other authors, he struggles to make life with Alzheimer's somehow meaningful and finds that meaning in narrating his experience of illness for others. "I would like to be somebody who could help understanding from the patient's point of view—what it is like to be an Alzheimer's patient," he says. "It's somebody's version of hell and I guess I'll someday have to write a book about that, which is exactly what I am trying to do. To whoever can read it to you, if you need to be read to, I would be glad to sell you a copy of my little Bible of Alzheimer's."[73] Like the other authors, he is still able to take comfort and pleasure in some aspects of life—music, for example, or the company of his small dog, Uni. If Henderson cannot create the same sort of narrative resolution as that of the other three, there remains an impression in many of his passages that he has come to a kind of peace with his life as an onlooker and philosopher. His editors conclude the book with such a passage: "As I ramble around there are some things that probably ought to be said that I'm sure I haven't said. But my philosophizing for this day is about finished. I can't imagine there's enough tape in there for any more. And what's more, I think I've said all that I can say. . . . The one thing I know is the dog is with me and when the dog is with me I at least have some solace, even if I don't know the way."[74]

In order to continue to understand themselves as intact people, each of these authors must radically redefine what it means for him or her to be a person. They can no longer base their sense of self strictly on the values of independence, self-control, and self-fulfillment. Confronting Alzheimer's, they must also find ways to value, or at least reconcile themselves to, dependency, contingency, and loss. In this way, their narratives can be seen to embody the tension between "modern" and "postmodern" models of self.

These narratives cannot, of course, be taken as empirical evidence for deciding the theoretical issue of whether people with Alzheimer's possess an intact self or whether and at what stage they lose it. While these authors clearly display an intact self, it could be argued that they are simply not "far enough gone." The losses they describe so vividly can be taken as evidence that their selfhood is in fact eroding. The accounts of both Sabat and Harré and Fontana and Smith are internally consistent and satisfying on their own terms, and commitment to one model or another thus ultimately hinges on political and moral grounds. What kind of social interactions do these models foster between people with dementia and those around them? What sort of problem does dementia become? What meanings do the models generate? Ultimately, what sort of world do they presume?

Fontana and Smith present their model as objective and do not frame it as a call for a specific policy initiative or change in the care of people with dementia, and this is also true in Hava Golander and Aviad Raz, the only other social constructionist account I am aware of that supports the idea of a loss of self in dementia.[75] This model could, however, be reasonably read as formal theorization of the "loss of self" theme of the Alzheimer's disease movement and the politics of anguish, thus supporting the massive public commitment to biomedical research as a way to address the problem of Alzheimer's. If the self is truly lost in Alzheimer's, then money would be better spent on research for prevention and cure rather than on treatment. In any case, the perspective of people with Alzheimer's disease need not in their account—indeed cannot—be considered.

By contrast, Sabat and Harré clearly position their model as a challenge to the prevailing idea that selfhood is lost in dementia and a call for improvement in the approach to caring for people with dementia. "What would it mean to the A.D. sufferer if the loss of [social] selves could be prevented?" they asked. "It would mean a lessening of social isolation, the continuation of relationships and the respect that a lifetime of living demands. It would mean that the only prison in which the sufferer would dwell would be that created by the boundaries of brain injury and not, in addition, the confinement that is brought about by the innocently misguided positioning and story lines created by others."[76] Improving the care of people with dementia was thus a moral imperative, as Sabat made clear in a later article: "We, the healthy, are obliged to join the battle and do our part to help the afflicted. In working to preserve the selfhood of the afflicted, we act in accordance with Kant's dictum that we should treat others as ends in themselves."[77]

Other authors who shared Sabat and Harré's commitment to reconstructing the self of the person with Alzheimer's expressed more concrete political commitments, positioning their work as a challenge to the hegemony of biomedicine and the "health politics of anguish." Karen Lyman, in an oft cited critique of the biomedical model of dementia, was willing to grant much to biomedicine. "Clearly, dementing illnesses involve disease processes for which biomedical research may hold the key to an eventual cure," she acknowledged. "But while awaiting a cure, care occurs in social settings and relationships that are seldom examined in regard to their contribution to dementia. . . . Reliance upon the biomedical model to explain the experience of dementing illness overlooks the social construction of dementia and the impact of treatment contexts and caregiving relationships on disease progression." Lyman granted that there were benefits to the "health politics of anguish." The "widespread acceptance of the biomedical model of dementia has countered the longstanding ageist assumption that senility is an inevitable condition of old age, has legitimized research and policy interests that offer hope for a cure, and has offered caregivers some degree of order and control in the difficult work of dementia care." But "in reality, the psychosocial experience of a dementing illness cannot be contained within biomedical concepts of brain disease," and the result was that the crucial perspective of the patient has been missing.[78] The implications of this challenge for policy were obvious. As psychiatrist and bioethicist Edmund G. Howe put it in a special issue of the *Journal of Clinical Ethics* devoted to the perspective of patients with dementia, these authors "argue that careproviders and others can arrest and sometimes, even, reverse these patients' downward course by interacting with them in a way that evokes positive emotions. If they are right, this insight is revolutionary, and raises profound new ethical questions. Among the most important of these questions, no doubt, is the extent to which society should act now in light of this awareness to give these patients the treatment they need."[79]

British psychologist Tom Kitwood, perhaps the most prominent and outspoken challenger of what he termed the "neuropathic ideology" of the biomedical approach to dementia, suggested that nothing less than a "cultural transformation" was involved in viewing people with dementia as fully human. More than anyone else, Kitwood operationalized the critique of the biomedical view of dementia as a brain disease that destroyed the selfhood of a person, developing not only a critique but a well-developed theory and method of caregiving that sustained the selfhood of those with dementia. Kitwood and his colleagues had successfully implemented this approach in small-scale institutional settings and in the community, but he acknowledged that there were many material and psychological barriers to providing the resources necessary to develop it on a large scale. These ranged, he argued, from the vested interests of the medical profession, pharmaceutical industry, and the state in the biomedical model to the psychological resistance to embracing people with dementia as fully human with which both professional and family caregivers struggled. But Kitwood believed that it was possible to overcome these obstacles by incorporating the values and practices of "person-centered" dementia care into existing structures, and the energy and enthusiasm thus generated could be used to bring about the large-scale institutional and political changes that were necessary: "The positive transformation of care practice—if it occurs on a widespread scale—will undermine all facile forms of determinism, and interrupt the obsessional search for technical fixes to human problems. It will challenge the stupidity and narrowness of the market mentality, and in particular the idea that human services can be effectively delivered as if they were consumer durables. It will stimulate the search for an economic process better than the one we have at present—a process which distributes the benefits of social existence far more equitably in relation to real need."[80]

Of course, "person-centered" dementia care was not entirely new. As described in chapters 2 and 3, an earlier generation of American psychiatrists and gerontologists had rejected the biomedical model of dementia as too reductionist, developed a theory of pathological social relations as a cause of dementia, and insisted that people with dementia retained an intact, if very challenged, selfhood that could be reached through appropriate therapeutic social interactions. Moreover, the success of what I have called the "gerontological persuasion" laid the political and scientific groundwork for the emergence of Alzheimer's disease as a major public issue.[81]

Yet in one important respect this theoretical work, whether entirely new or not, would, if put into practice, truly constitute a cultural transformation. Far more than the previous generation of psychiatrists and gerontologists, scholars such as Sabat, Harré, Lyman, and Kitwood were intent on erasing the boundaries between the normal and the demented. The earlier generation of researchers had assumed that people with dementia were still people and that, as people, they could benefit therapeutically from positive social interactions. Such a view is perhaps preferable to seeing them as "empty shells," but as we have seen, it left plenty of room for the continued stigmatization of both people with dementia and old age in general. Contemporary scholars and researchers working from the social constructionist position go beyond this to claim that people with dementia are in fact not different in any essential or pathological way from any other people and that some of the qualities of people with dementia were actually positive. If, as Foucault suggests in *Madness and Civilization,* a central characteristic of modernity has been the exclusion of the mentally ill, the refusal to allow their voice to enter into our discourse without the mediating structure of disciplinary knowledge, then embracing the radical egalitarianism of person-centered dementia care would indeed be a cultural transformation.[82]

Kitwood expressed this stance most forcefully. He and colleague Kathleen Bredin published an article arguing that it was necessary to see that "we" who are normal as much as "they" who are cognitively impaired are part of the problem of dementia:

In some respects, indeed, *we* might be considered more problematic than *they*. *They* are obviously damaged, derailed, deficient: a neurological process has interfered with their everyday functioning. But *we*, in our varied ways, are damaged, derailed, deficient too. What is particularly dangerous here is that generally *we* do not acknowledge these facts. Indeed, much of our professional socialization involves a systematic training in how to avoid such a painful insight. . . . What is more, it is arguable that the general pattern of everyday life, with its hypocrisy, competitiveness and pursuit of crass materialism is, from a human standpoint, deeply pathological; those whose way of being dovetails smoothly into this pattern are the most "normal" and well-adjusted. The truth, from this other standpoint, is that such people simply participate in an unacknowledged "pathology of normality." Thus, when the interpersonal field surrounding the beginnings of "dementia" is looked at in this way, the problem is by no means focused on a single person whose brain is failing. Those others who have face-to-face contact are also involved; and in the background, so also is the prevailing pattern of social relations.[83]

In his book, Kitwood expanded on this idea. "Contact with dementia or other forms of severe cognitive disability can—and indeed should—take us out of our customary patterns of over-busyness, hypercognitivism and extreme talkativity, into a way of being in which emotion and feeling are given a much larger place." People with dementia have, according to Kitwood, important things to teach the rest of humankind. More than anything else, they force us to reconsider what it means to be a person. Individuality and autonomy are radically questioned, and the interdependence of all people becomes obvious. "Frailty, finitude, dying, death are rendered more acceptable; grandiose hopes for technical Utopias are cut to the ground. Reason is taken off the pedestal that it has occupied so unjustifiably, and for so long; we reclaim our nature as sentient social beings. Thus from what might have seemed the most unlikely quarter, there may yet emerge a well-spring of energy and compassion."[84]

In living lives that are of necessity free of the shackles of rationality, stable narratives, the illusions of independent agency, and the unity of experience, the person with Alzheimer's in social constructionist discourse becomes a veritable prophet of the postmodern self.[85] If illness functions as a cultural metaphor, then a disease that threatens the foundations of modern notions of selfhood would seem the perfect metaphor for an age in which technology wildly multiplies human relationships to the point of "social saturation," with the result that "the very concept of personal essences is thrown into doubt. Selves as possessors of real and identifiable characteristics—such as rationality, emotion, inspiration, and will—are dismantled."[86] A disease whose most prominent feature is the destruction of memory, and whose most dreaded moment is when victims no longer recognize friends and family members they have known for a lifetime, seems to encapsulate the erosion of self in a culture in which a selfhood rooted in personal history, friends, family, and a sense of place becomes increasingly difficult to maintain because "selective apathy, emotional disengagement from others, renunciation of the past and the future, a determination to live one day at a time . . . have come to shape the lives of ordinary people under the ordinary conditions of a bureaucratic society widely perceived as a far-flung system of total control."[87] Dementia, it would seem, is a problem with which we all struggle.

Certainly, there are times when the published narratives of people with Alzheimer's disease seem to speak to more general problems of personhood, as when Diana Friel McGowin writes that "each one of us must feel they have worth as a living being—a person with the same rights and privileges as the people next door. I feel my lack of worth acutely when I am in large groups of people. Being in a crowd or even in a busy thoroughfare overwhelms me. All those people 'of worth,' with places to go—who know where they are going."[88] Similarly, Cary Smith Henderson could be speaking of the general problem of the fragmentation and randomness of contemporary life when he says that "no two days and no two moments are the same. You can't build on experience. You can maybe guess what's going to happen a little while from now—minutes from now, hours from now—we don't know what to expect."[89]

Whatever the merit of this argument—and I believe there is much—it is important to give some credence to Susan Sontag's warning that our propensity to use illness as metaphor results in a romanticization of disease that adds to the stigma of the disease and obscures the real experience of illness.[90] As we have seen, stereotypes have positive as well as negative aspects, and the image of the person with Alzheimer's as postmodern prophet may be driven by the same basic anxieties about selfhood as the image of the Alzheimer's victim as a hollow shell.[91] In attempting to deemphasize the boundaries between the normal and the pathological, caregivers and sympathetic scholars need to remember that those suffering from Alzheimer's face very real problems that most people do not. Thus, for example, the well-intentioned claim that "we are all demented" may contain some important truths, but is also in an obvious way a cruel lie. Larry Rose, with early-onset Alzheimer's disease and in the early stages of dementia, expressed great irritation at well-meaning people who, when he responded to their questions about the symptoms he had, tried to reassure him by telling him that they did such things all the time. He recounted an incident in which a friend called him up, describing a hectic day at work in which he had made several mistakes and telling Rose, "If *you* have Alzheimer's, I must have a *double* dose of it." Rose describes his angry reaction:

> "Do you forget simple words, or substitute inappropriate words, making your sentence incomprehensible? Do you cook a meal and not only forget you cooked it, but forget to eat it? Do you put your frying pan in the freezer, or your wallet in the sugar bowl, only to find them later and wonder what in the world is happening to you? Do you get lost on your own street . . . ? Do you become confused or fearful ten times a day, for no reason? And most of all, do you become irate when someone makes a dumb statement like you just made? . . . Then you don't have Alzheimer's disease," I said, and hung up.[92]

Rose regretted his harshness toward his friend, but his point needs to be taken seriously. We should perhaps distinguish between the dementia produced by the hypocognitive situation of the person with Alzheimer's and the dementia produced by a hypercognitive society. Both are profoundly, perhaps at times even equally, disorienting and disruptive of a coherent sense of self. But there is a difference between having one's cognitive abilities impaired to the degree that one cannot successfully perform expected social roles and experiencing confusion—even extreme confusion—because social roles that one successfully performs are contradictory and incoherent.

But this does not mean that we can or should back away from the political and moral implications that social constructionist accounts of dementia have raised. As many scholars have pointed out, Sontag is wrong in thinking that we can somehow get beyond using illness as metaphor to produce a pristinely objective scientific discourse, free of the representational web of human language.[93] Medical science at its most objective is itself a set of values revolving around rationality and the desire for technical mastery of the world, and its practices serve ideological purposes. For example, a number of scholars have shown that the process of diagnosis itself—however objective the clinical and technological procedures appear to be— is best understood as efforts to organize human experience according to these values.[94] Citing the work of physician-sociologist Howard Waitzkin, Arthur Frank argues that medicine can be understood "as an ideological system that 'calls' the patient to be an identity that medicine maintains for him; the diagnosis is the most prevalent form of this identity. The ideological work of medicine is to get the patient to accept this diagnostic identity as appropriate and moral. When the patient accepts this identity, he aligns himself as subordinate in a power relation."[95]

From this perspective, the moral problem of illness is not that meanings are assigned to what should be a value-free realm of experience but that the meanings medicine assigns to illness obscure the experiences and silences the voices of the people who are ill. This is perhaps the most important aspect of the critique of the rise to prominence of Alzheimer's disease in American culture. While the biomedical model of Alzheimer's disease created opportunities for a richer understanding of the pathological processes within the brain that are associated with dementia, it was connected to a discourse on policy that pitted increased funds for research against increased funds for caregiving, making it harder to solve the problem of long-term health insurance. Moreover, this discourse created a meaning for Alzheimer's disease which had devastating implications for people who have it—calling into question their very moral status as persons. This is not to say that biomedical research should not be supported, but we must find a way of balancing the imperatives of one fruitful scientific construct of Alzheimer's disease (among others that might be equally fruitful) with other, equally important, political and moral imperatives.

The discourses of caregivers, professionals, and people with dementia themselves provide grounds for optimism that a cultural space is being created in which these different imperatives can be held in creative tension. But as the molecular biology of Alzheimer's becomes better understood, there may be a tendency to focus all attention and hope on the development of a molecular miracle drug. We must take care that our understandable drive to find a prevention or cure for the disease does not blind us to the need to provide attention and resources to the social and moral complexities that dementia entails. We must never lose sight of the imperative to reach out to the humanity of those struggling with dementia and to be open to the challenge that their experience may pose to our ideas of what it means to be a person.

Notes

1. Robert Gard, *Beyond the Thin Line* (Madison, Wis.: Prairie Oak Press, 1992), pp. ix–x.

2. Ibid., pp. 21–23.

3. Peter J. Whitehouse, Konrad Maurer, and Jesse F. Ballenger, eds., *Concepts of Alzheimer Disease: Biological, Clinical, and Cultural Perspectives* (Baltimore: Johns Hopkins University Press, 2000), p. xi.

4. Benjamin Rush, *An Account of the Causes and Indications of Longevity, and of the State of the Body and Mind in Old Age; with Observations on its Diseases, and their Remedies* (1793), reprinted as "On Old Age," in Dagobert V. Runes, *Selected Writings of Benjamin Rush* (New York: Philosophical Press, 1947); quotations are from Runes, pp. 349, 351; emphasis in original.

5. The history of aging has focused primarily on New England. The situation is much less clear for other regions of the country.

6. Thomas Cole, *The Journey of Life: A Cultural History of Aging in America* (Cambridge: Cambridge University Press, 1992), chaps. 3–4; Carole Haber and Brian Gratton, *Old Age and the Search for Security: An American Social History* (Bloomington: Indiana University Press, 1994), pp. 144–165.

7. The minister was Nathaniel Emmons. His career and this sermon are described by Cole in *Journey of Life*, pp. 57–66.

8. Ibid., pp. 103–104. Haber and Gratton (*Old Age and the Search for Security*) do not explicitly discuss the effect of Enlightenment ideals on representations of old age, though their discussion of the correspondence between Jefferson and Adams late in their lives regarding old age tends to support Cole's view. Cole disagrees with the early interpretation of David Hackett Fischer (*Growing Old in America* [Oxford: Oxford University Press, 1978]), who argued that the association of old age with hierarchy and privilege which this religious model entailed was shattered in the ideological ferment of the American Revolution. According to Cole, Fischer was not wrong in suggesting that Enlightenment ideals and the democratic spirit of the revolutionary era tended to work against the traditional view of age, but Cole suggests that this was a much more gradual process. Cole, *Journey of Life*, p. 56, n. 41.

9. In *Growing Old in America* (pp. 60–67), Fischer described many exceptions to the rule of veneration, which applied most unequivocally to the wealthy and powerful men who

dominated society and "probably existed within most families where the husband and father was healthy and active" (p. 60). The aged poor, especially widows, and aged slaves were more likely to be scorned than venerated. Throughout their *Old Age and the Search for Security*, Haber and Gratton are admirably sensitive to the way in which race, class, and gender structured the experience of aging throughout American history.

10. Quoted in Haber and Gratton, *Old Age and the Search for Security*, p. 147.

11. Benjamin Rush, *The Autobiography of Benjamin Rush: His "Travels through Life" together with His Commonplace Book for 1789–1813* (Princeton, N.J.: American Philosophical Society, 1948), pp. 215–216.

12. Haber and Gratton's interpretation of Rush's text is somewhat at odds with that of Cole, which I am essentially elaborating on here. Primarily concerned with the issue of the status of the elderly rather than the meaning of old age, Haber and Gratton are more impressed with Rush's prescription for a passive old age than with his connections to Enlightenment and religious ideals that found meaning in the suffering and loss of old age. Thus, they suggest that "his view of old age was sentimental at best; in its romanticization, the ideal old age lost power and importance." Rush's prescription for old age, like that of many other writers and ministers, was passivity. "Although they possessed no visible infirmities, old men and women would simply come to accept their fate and focus serenely on the passing of life" (Haber and Gratton, *Old Age and the Search for Security*, p. 154). Thus, in Haber and Gratton's account, Rush was less connected to the tradition of veneration for the aged than to the emergence of ageism in the nineteenth century. The difference in these interpretations is essentially one of emphasis. If the issue is the meaning of old age, Rush's vision of older people as exemplars of patient Christian suffering is clearly connected to the meaning of old age articulated by the New England clerics.

13. Rush, "On Old Age," pp. 356–357.

14. Rush was not denying that the elderly were hard of hearing; rather, he was explaining that when aural stimuli are received, they more quickly excite "sensation and reflection" than visual stimulation does. Thus Rush reports that "Mr. Hutton informed me, that he has frequently met his sons in the street without knowing them until they had spoken to him. [And] Dr. Franklin informed me that he recognized his friends, after a long absence from them, first by their voices" (ibid., p. 347).

15. Ibid., p. 349.

16. Ibid., p. 350. Swift is a particularly interesting figure because he has been"diagnosed" as fairly clearly having had Alzheimer's disease by a number of "historically minded" Alzheimer's disease researchers. Swift is also credited with presenting an early, nonprofessional yet highly accurate description of the disease in his "struldbrugs" (the people who never died in *Gulliver's Travels*) and with supporting a facility for the humane treatment of people with dementia.

17. Ibid., p. 350.

18. Ibid., pp. 350–351.

19. For discussions of some of the controversial issues concerning dementia prevalence rates, see K. G. Losonczy, L. R. White, and D. B. Brock, "Prevalence and Correlates of Dementia: Survey of the Last Days of Life," *Public Health Reports* 113, no. 3 (1998): 273–280, and V. S. Thomas et al., "Estimating the Prevalence of Dementia in Elderly People: A

Comparison of the Canadian Study of Health and Aging and National Population Health Survey Approaches," *Psychogeriatrics* 13, suppl. 1 (2001): 169–175.

20. Alzheimer's Association, *Fact Sheet: Alzheimer's Disease Statistics* (Chicago: Alzheimer's Association, 2004).

21. Daniel Walker Howe, *Making the American Self: Jonathan Edwards to Abraham Lincoln* (Cambridge: Harvard University Press, 1997); Wilfred M. McClay, *The Masterless: Self and Society in Modern America* (Chapel Hill: University of North Carolina Press, 1994).

22. See McClay, *Masterless*, and T. J. Jackson Lears, *No Place of Grace: Antimodernism and the Transformation of American Culture, 1880–1920* (New York: Pantheon, 1981), and "The Ad Man and the Grand Inquisitor: Intimacy, Publicity, and the Managed Self in America, 1880–1940," in *Constructions of the Self,* ed. George Levine (New Brunswick, N.J.: Rutgers University Press, 1992).

23. Julie Hilden, *The Bad Daughter* (Chapel Hill, N.C.: Algonquin, 1998), p. 152.

CHAPTER 1: THE STEREOTYPE OF SENILITY IN
LATE-NINETEENTH-CENTURY AMERICA

1. William Osler, "The Fixed Period," in *Aequanimitas*, 3rd ed. (Philadelphia: Blakiston, 1947), pp. 382–383.

2. Michael Bliss, *William Osler: A Life in Medicine* (Oxford: Oxford University Press, 1999), pp. 312–314.

3. Harvey Cushing, *The Life of Sir William Osler*, vol. 1 (Oxford: Oxford University Press, 1926), pp. 665–670.

4. For descriptions of how the controversy developed, see Bliss, *William Osler,* pp. 324–325, and Cushing, *Life of Sir William Osler,* pp. 668–669.

5. Osler's published statement and letter in the *New York Times* are reproduced in Charles F. Wooley, "William Osler and the *New York Times*, 1897–1931," in *The Persisting Osler-III: Selected Transactions of the American Osler Society, 1991–2000,* ed. Jeremiah A. Barondess and Charles G. Roland (Malabar, Fla.: Krieger, 2002), p. 53.

6. Osler, *Aequanimitas,* p. viii.

7. William Graebner, *A History of Retirement: The Meaning and Function of an American Institution, 1885–1978* (New Haven, Conn.: Yale University Press, 1980), pp. 3–10. Graebner argued that the criticism directed at Osler was not—as Harvey Cushing, Osler's friend, longtime associate, and biographer, contended—a result of a credulous public taking the chloroform remark seriously. Most newspapers made it clear enough that it was a joke. Rather, the controversy was a response to the threatening implications of Osler's assessment of the relative worth of older men. Articles critical of the address typically included lists of people who had made signal contributions to civilization in middle and old age. Most other historians of aging have followed Graebner's analysis of the incident. Bliss's account (*William Osler,* pp. 312–314) adds an understanding of Osler's feelings toward the remarks and his continued insistence on the basic point that older people contributed little productive work to society.

8. Major works in the history of old age have argued different periodization schemes for the denigration of old age. David Hackett Fischer's early foray into the field (*Growing Old*

in America [Oxford: Oxford University Press, 1978]) claimed there was a "revolution in age relations" from 1770 to 1820 in which the egalitarian ideology of the American Revolution and Enlightenment ideals of rationality and human agency led to a widespread attack on the veneration and social status of the aged as yet another example of arbitrary authority. Subsequent works have rejected this scheme, emphasizing that this change was more gradual and centered in the second half of the nineteenth century. W. Andrew Achenbaum (*Old Age in the New Land: The American Experience since 1790* [Baltimore: Johns Hopkins University Press, 1978]) argued that such a change occurred between the Civil War and World War I but had more to do with the belief that the elderly were ill equipped to cope with the increasingly rapid pace of life in an industrial society. Carole Haber (*Beyond Sixty-Five: The Dilemma of Old Age in America's Past* [Cambridge: Cambridge University Press, 1983]) made a similar argument, focusing especially on the development of negative assumptions about aging and old age by physicians and other professionals, whose authority and prestige were increasing in the last third of the nineteenth century. Thomas Cole (*The Journey of Life: A Cultural History of Aging in America* [Cambridge: Cambridge University Press, 1992]) locates this change in the attitudes of evangelical reformers in the antebellum period but emphasizes the gradual nature of the change and that it was not fully accomplished until late in the century. Carole Haber and Brian Gratton's synthesis of historical work on old age (*Old Age and the Search for Security: An American Social History* [Bloomington: Indiana University Press, 1994]) suggests that all aspects of the history of old age can be usefully periodized in terms of the preindustrial, industrial, and social security eras but argues that change was complex. Although old age was increasingly denigrated by experts and in popular discourse in the industrial era, Haber and Gratton have produced data on the labor force participation and incomes of elderly people which suggest that the industrial era brought significant improvement to the lives of many of them who were able to continue work in relatively high paying industrial jobs.

9. Achenbaum first noted this change in meaning in his *Old Age in the New Land*. But Haber, with greater emphasis on the history of medicine, develops the idea much more fully and supplies the Jefferson quote on p. 74 of her *Beyond Sixty-Five*. Moving beyond the concept of a simple change in usage, she shows that the term reflected the growing pessimism among American physicians about old age in the last quarter of the nineteenth century as they absorbed a more elaborate knowledge of European neuropsychiatry and pathology. Postmortem findings were revealing that the brain—the organ of the mental and moral faculties—shrunk, thinned, and became riddled with lesions as it aged, and these findings eroded the sense that intellect and spirit could somehow transcend the decay of the body. Haber develops this line of thought in chapters 3 and 4 of *Beyond Sixty-Five* and in her essay "Geriatrics: A Specialty in Search of Specialists," in *Old Age in a Bureaucratic Society: The Elderly, the Experts and the State in American History*, ed. David D. Van Tassel and Peter N. Stearns (Westport, Conn.: Greenwood Press, 1986).

10. Cole, *Journey of Life*, pp. 56, 93.

11. The minister was the conservative Cortland Van Rensselaer, quoted in ibid., p. 86.

12. Haber and Gratton, *Old Age and the Search for Security*, pp. 155–163. See also Cole, *Journey of Life*, chap. 5.

13. Richard H. Shryock, *Medicine and Society in America: 1660–1860* (New York: New

York University Press, 1960); Paul Starr, *The Social Transformation of American Medicine* (New York: Basic Books, 1982), chaps. 1, 3. On the importance of training in France and Germany to American medical students, see chapter 19 of Erwin Ackerknecht's *A Short History of Medicine*, rev. ed. (Baltimore: Johns Hopkins University Press, 1982).

14. On the metaphoric extension of the laws of thermodynamics to a broad spectrum of disciplines concerned with the workings of the human body, see Anson Rabinbach, *The Human Motor: Energy, Fatigue, and the Origins of Modernity* (Berkeley: University of California Press, 1990).

15. Haber, *Beyond Sixty-Five*, pp. 50–51, 65; Haber, "Geriatrics," pp. 79–80.

16. Haber, *Beyond Sixty-Five*, p. 65.

17. Charles Mercier, *Sanity and Insanity*, 2nd ed. (New York: Macmillan, 1908), p. 371.

18. Ibid.

19. Such a delay actually is one of the aims of therapeutic interventions into Alzheimer's disease that emerged in the 1990s. Few responsible voices believe a cure or outright prevention of Alzheimer's disease will be available anytime soon. But it does seem reasonable to expect that improvements in the therapies currently available and in development will be able to slow the progress of dementia to some degree—ideally to the point where serious dementia would be a relatively rare occurrence in the extremely aged. Of course, gerontological imperatives are somewhat circular, for other gerontologists boldly predict that the normal human life span will soon be increased to 120 or more years, which could counteract the positive effects we can expect from drug therapies in delaying the onset of Alzheimer's disease. See, for example, Roy L. Walford. *Maximum Life Span* (New York: Norton, 1983).

20. Ralph Lyman Parsons, "Practical Points Regarding the Senile Insanities with Special Reference to Prophylaxia and Management," *Medical Record* 50 (1896): 505–508.

21. William Alcott, *The Laws of Health* (Boston: J. P. Jewett, 1860), quoted in Cole, *Journey of Life*, p. 105. On Alcott's career in the context of popular health reform in the nineteenth century, see chapter 5 of Cole.

22. Parsons, "Practical Points," p. 505.

23. H. Wardner, "Extract from a Report on the Disease of Old Age," *Alienist and Neurologist* 4 (1893): 103.

24. Carole Haber, "From Senescence to Senility: The Transformation of Old Age in the Nineteenth Century," *International Journal of Aging and Human Development* 19 (1984): 41–45.

25. W. Bevan Lewis, *A Textbook of Mental Diseases: With Special Reference to the Pathological Aspects of Insanity* (Philadelphia: Blakiston, 1890), cited in William J. Russell, "Senility and Senile Dementia," *American Journal of Insanity* 63 (1902): 626.

26. Russell, "Senility and Senile Dementia," 628.

27. Michael Shepherd, "Psychiatric Journals and the Evolution of Psychological Medicine," in *Medical Journals and Medical Knowledge: Historical Essays*, ed. W. F. Bynum, Stephen Lock, and Roy Porter (New York: Routledge, 1992).

28. Of the fourteen articles concerning senile dementia, cerebral arteriosclerosis, and Alzheimer's disease which appeared in these two journals during this period, all but four were primarily concerned with describing the pathological structures of these conditions.

Of the four, two were concerned with differential diagnosis and another with hypertension as an indicator of cerebral arteriosclerosis. The final article was an interesting non-Freudian discussion of a patient's recurrent dream as a precursor to dementia.

29. For examples of this kind of article, see C. G. McGaffin, "An Anatomical Analysis of Seventy Cases of Senile Dementia," *American Journal of Insanity* 66 (1910): 649–673, and William J. Tiffany, "The Occurrence of Miliary Plaques in Senile Brains," *American Journal of Insanity* 70 (1914): 695–739.

30. E. E. Southard, "Anatomical Findings in Senile Dementia: A Diagnostic Study Bearing Especially on the Group of Cerebral Atrophies," *American Journal of Insanity* 66 (1910): 673.

31. Though they disagree on most interpretive issues regarding this development, both David J. Rothman and Gerald Grob agree on the relatively high prestige of psychiatry through the mid-nineteenth century and the importance of the mental hospitals to that prestige. See David J. Rothman, *The Discovery of the Asylum: Social Order and Disorder in the Early Republic* (Glenview, Ill.: Scott, Foresman, 1971), and Gerald Grob, *Mental Institutions in America: Social Policy to 1875* (New York: Free Press, 1973) and *Mental Illness and American Society, 1875–1940* (Princeton, N.J.: Princeton University Press, 1983).

32. The cure rates reported by antebellum psychiatrists are not entirely reliable, as Grob has pointed out. The high rates of successful treatment claimed by asylum superintendents in the 1830s and 1840s were questioned by psychiatrists later in the century: these statistics failed to take into account readmission of patients who had been discharged as recovered and thus counted individuals who were "cured" more than once. On the other hand, some follow-up studies that addressed these criticisms indicated that the hospitals enjoyed considerable success—finding, for example, that as many as 51% of the patients discharged from a state hospital prior to 1840 were never admitted again and so presumably had recovered well enough to live the rest of their lives in the community. Grob, *Mental Institutions in America*, pp. 182–185.

33. F. H. Stephenson, "Senility, Senile Dementia and Their Medico-Legal Aspects," *Buffalo Medical Journal* 56 (1901): 559.

34. See Charles Rosenberg, "George M. Beard and American Nervousness," in *No Other Gods: On Science and American Social Thought*, revised and expanded ed. (Baltimore: Johns Hopkins University Press, 1997); Barbara Sicherman, "The Uses of a Diagnosis: Doctors, Patients, and Neurasthenia," *Journal of the History of Medicine and Allied Sciences* 32 (1977): 33–54; Francis G. Gosling, *Before Freud: Neurasthenia and the American Medical Community, 1870–1910* (Chicago: University of Illinois Press, 1987); and Tom Lutz, *American Nervousness, 1903: An Anecdotal History* (Ithaca, N.Y.: Cornell University Press, 1991).

35. Rosenberg, "George M. Beard," p. 98.

36. Achenbaum, *Old Age in the New Land*, pp. 46–47; Cole, *Journey of Life*, chap. 8; Graebner, *History of Retirement*, p. 30.

37. Beard's inclusive language in this latter quotation was an anomaly. No woman appeared in the dozens of concrete examples he lists, though they are figured into a graph he provides to illustrate his "law"; neither did a woman appear in his discussion of legal responsibility. His only mention of women is an abstract observation suggesting that the timing of menopause corresponds with the decline of mental powers. George M. Beard,

Legal Responsibility in Old Age: Based on Researches into the Relation of Age to Work (1874), reprinted in *The "Fixed Period" Controversy*, ed. Gerald Gruman (New York: Arno Press, 1979).

38. Beard, *Legal Responsibility*, pp. 7–8, 11 (all page numbers are from the volume edited by Gruman, cited in the previous note); emphasis in original.

39. Ibid., pp. 11, 33, 36.

40. Ibid., footnote on p. 11.

41. Ibid., pp. 19–20, footnote on p. 11, p. 30; emphasis in original.

42. For a discussion of Nascher's importance to geriatrics and gerontology, see W. Andrew Achenbaum, *Crossing Frontiers: Gerontology Emerges as a Science* (New York: Cambridge University Press, 1995), pp. 43–45.

43. Ignatz Leo Nascher, *Geriatrics: The Diseases of Old Age and Their Treatment* (1914; facsimile reprint, New York: Arno Press, 1979), p. 12.

44. Ibid., pp. v–vi.

45. Ibid., p. 16.

46. Ibid., p. 483.

47. Ibid., pp. 504–506.

48. Ibid., p. 138.

49. Beard, *Legal Responsibility*, p. 9.

50. Osler, "Fixed Period," p. 383.

51. Lucy Martin Donnelly, "In Praise of Old Ladies," *Atlantic* 84 (1899): 852.

52. "Passing of the Old Lady," *Atlantic* 99 (1907): 874.

53. "A Word for the Modern Old Lady," *Atlantic* 100 (1907): 283–284.

54. Nascher, *Geriatrics*, pp. 16–17.

55. Ignatz Leo Nascher, "Senile Mentality," *International Clinics*, 2nd ser., 4 (1911): 52.

56. Ignatz Leo Nascher, "Evidences of Senile Mental Impairment," *American Journal of Clinical Medicine* 22 (1915): 544.

57. Beard, *Legal Responsibility*, p. 34.

58. Nascher, "Evidences of Senile Mental Impairment," pp. 544–545.

59. Sander L. Gilman, *Disease and Representation: Images of Illness from Madness to AIDS* (Ithaca, N.Y.: Cornell University Press, 1988), pp. 5–6.

60. This discourse is described by Haber and Gratton in chapter 5 of *Old Age and the Search for Security* and in Howard Chudacoff, *How Old Are You? Age Consciousness in American Culture* (Princeton, N.J.: Princeton University Press, 1989).

61. I have not found an example of an article in this period in which an author identified him- or herself as a member of a different ethnic group or even addressed the possibility of an audience containing people of different ethnicities.

62. Walt Mason, "I Refuse to Grow Old," *American Magazine* 88 (1919): 67.

63. Ibid.

64. Ibid.

65. Five articles by the same anonymous author appeared in *Harper's Bazaar* in 1906–7 under the title "The Land of Old Age," by an "Elderly Woman." As indicated by identical names for family members and situations, the same author apparently wrote "The Autobiography of an Elderly Woman" for *Everybody's Magazine* in 1906.

66. "Land of Old Age: Part IV. The Conventions of Age." *Harper's Bazaar* 41 (1907): 365–366.

67. "Land of Old Age: Part IV," pp. 366, 367–368.

68. "Autobiography of an Elderly Woman," *Everybody's Magazine* 14 (1906): 626.

69. "The Land of Old Age: Part V. Young People and Old." *Harper's Bazaar* 41 (1907): 1057.

70. "The Land of Old Age: Part V," p. 1059.

71. "Autobiography of an Elderly Woman," p. 628.

72. G. Stanley Hall, *Senescence: The Last Half of Life* (1922; facsimile reprint, New York: Arno Press and New York Times, 1972), pp. viii–ix.

73. Ibid., p. ix.

74. Cole, *Journey of Life*, chap. 10.

75. Gilman, *Disease and Representation*, pp. 6–7.

CHAPTER 2: BEYOND THE CHARACTERISTIC PLAQUES AND TANGLES

1. If one assumes that the prevalence of dementia in the population 65 and over was about the same in 1900 and 1930 as it is today—approximately 10%—one could make the crude estimate that there were at most about 300,000 people with dementia in 1900 and about 600,000 in 1930. But since the 85-and-over population—in which the prevalence of dementia runs closer to 50%—was a much smaller part of the overall elderly population at the beginning of the twentieth century than it is today, the prevalence in 1900 and 1930 was probably even lower.

2. On the abandonment of moral treatment as the theoretical basis for psychiatry and the psychiatric hospital, see Gerald Grob, *Mental Institutions in America: Social Policy to 1875* (New York: Free Press, 1973), pp. 182–185, and chapter 6 of Nancy Tomes, *A Generous Confidence: Thomas Story Kirkbride and the Art of Asylum-Keeping, 1840–1883* (Cambridge: Cambridge University Press, 1984). On the ambivalence of the psychiatric profession toward the state mental hospitals, see chapter 11 of Gerald Grob, *Mental Illness and American Society, 1875–1940* (Princeton, N.J.: Princeton University Press, 1983).

3. These journals were the *American Journal of Psychiatry*, the official journal of the American Psychiatric Association, and the *Archives of Neurology and Psychiatry*, the American Medical Association's official journal for these two specialties. In comparison, forty-six such articles appeared in the comparable journals for the decade 1971–80, when interest surged with the Alzheimer's disease movement. Journals for this decade were again the *American Journal of Psychiatry* and the journals *Archives of Neurology* and *Archives of General Psychiatry*, the latter two created when, in 1960, the American Medical Association split the *Archives of Neurology and Psychiatry* into separate journals, reflecting the growing autonomy of the two specialties.

On the split of the *Archives of Neurology and Psychiatry* into these two journals and its implications for the specialties of neurology and psychiatry, as well as the development of professional journals in the neurosciences in the United States and Great Britain, see Michael Shepherd, "Psychiatric Journals and the Evolution of Psychological Medicine," in

Medical Journals and Medical Knowledge, ed. W. F. Bynum, Stephen Lock, and Roy Porter (New York: Routledge, 1992).

4. Typically, historical accounts begin with Alzheimer's first description of the disease in 1906 and then jump to the research of the 1970s—implying that most of the twentieth century was essentially a dark age. For examples of this kind of account, see John F. Vannoy and James A. Greene, "Alzheimer's Disease: A Brief History," *Journal of the Tennessee Medical Association* 82 (January 1989); G. E. Berrios, "Alzheimer's Disease: A Conceptual History," *International Journal of Geriatric Psychiatry* 5 (1990): 355–365; and Patrick Fox, "Alzheimer's Disease: An Historical Overview," *American Journal of Alzheimer's Care* 1 (Fall 1986): 18–24. In his book on the history of research on the genetics of Alzheimer's (*Hannah's Heirs: The Quest for the Genetic Origins of Alzheimer's Disease*, expanded ed. [Oxford: Oxford University Press, 1996]), Daniel A. Pollen devotes one chapter to research on the disease at midcentury. Remarkably, he describes only *four* different articles, which he says are the only significant things written on the topic during this period, leaving the impression that these articles were the only ones that were published when, of course, scores of articles had appeared in the psychiatric literature on late-life dementias. The one major exception to the rule of ignoring midcentury psychiatry's concept of senile dementia is Martha Holstein's Ph.D. dissertation, "Negotiating Disease: Senile Dementia and Alzheimer's Disease, 1900–1980" (University of Texas at Galveston, 1996), which devotes a chapter to the psychodynamic model.

5. Carole Haber, *Beyond Sixty-Five: The Dilemma of Old Age in America's Past* (New York: Cambridge University Press, 1983), pp. 88–91; Gerald Grob, "Explaining Old Age History: The Need for Empiricism," in *Old Age in a Bureaucratic Society: The Elderly, the Experts and the State in American History*. ed. David D. Van Tassel and Peter N. Stearns (Westport, Conn.: Greenwood Press, 1986).

6. Grob, *Mental Illness and American Society*, pp. 182–184; p. 317 for discussion of the deinstitutionalization that began in the 1960s.

7. Charles G. Wagner, *N.Y State Commission on Lunacy, Annual Report 12* (1900), cited in Grob, *Mental Illness and American Society*, p. 186.

8. Grob, *Mental Illness and American Society*, p. 182 for New York, p. 184 for Massachusetts.

9. Richard W. Hutchings, "The President's Address," *American Journal of Psychiatry* 96 (1939): 4.

10. Harold W. Williams et al., "Studies in Senile and Arteriosclerotic Psychoses: Relative Significance of Extrinsic Factors in Their Development," *American Journal of Psychiatry* 98 (1942): 712.

11. Francis Braceland and Donald W. Hastings, "Somatopsychic Disorders of Old Age," *American Journal of Psychiatry* 99 (1942): 856.

12. Abraham Myerson, "Some Trends in Psychiatry," *American Journal of Psychiatry*, Special Centennial Issue (1944): 162.

13. See Grob, *Mental Illness in American Society*, p. 124, for a discussion of these efforts.

14. Gerald Grob describes these in *From Asylum to Community: Mental Health Policy in Modern America* (Princeton, N.J.: Princeton University Press, 1991), pp. 73–76.

15. Gunther E. Wolff, "Results of Four Years Active Therapy for Chronic Mental Patients and the Value of and Individual Maintenance Dose of ECT," *American Journal of Psychiatry* 114 (1957): 453.

16. For example, see Howard Rusk, "America's Number One Problem—Chronic Disease in an Aging Population," *American Journal of Psychiatry* 106 (1949): 270, and Lawrence Kolb, "The Mental Hospitalization of the Aged: Is It Being Overdone?" *American Journal of Psychiatry* 112 (1956): 627.

17. Grob, *From Asylum to Community*; see chapter 10 for policy changes regarding deinstitutionalization, pp. 267–270 for the role of Medicare/Medicaid.

18. More broadly, one could argue that psychoanalysis and psychodynamic psychiatry emerged in similar circumstances. Freud began his medical career as a neurophysiologist and never gave up the idea that all psychological illnesses would eventually be explained as disturbances of brain structure or function and amenable to somatic treatment. But because neuroanatomy and neurophysiology had so far yielded so little in elucidating the neuroses and functional psychoses, he resolved to confine his work to the psychological level. Erwin H. Ackerknecht, *A Short History of Psychiatry* (New York: Hafner, 1959), pp. 81–82. Interestingly, Edward Shorter's recent *History of Psychiatry: From the Era of the Asylum to the Age of Prozac* (New York: John Wiley and Sons, 1997) emphasizes a similar dynamic in the development of a modern classification scheme by Emil Kraepelin. According to Shorter, Kraepelin had turned away from biological theories of causation because they were simply too speculative. Until etiological theories of any sort could be better substantiated, psychiatric classification should rest on clinical observation of the natural history of disease (p. 106).

19. On American psychiatry's drive to ally itself with these other branches' use of pathology and bacteriology in the last two decades of the nineteenth century, see chapter 3 of Grob's *Mental Illness and American Society*. Also see Charles Rosenberg's argument that psychiatry's inability to identify discrete disease categories with a clear pathological basis has been a determining characteristic in its history through the twentieth century: Charles Rosenberg, "The Crisis in Psychiatric Legitimacy: Reflections on Psychiatry, Medicine, and Public Policy," *Explaining Epidemics and Other Studies in the History of Medicine* (New York: Cambridge University Press, 1992). Another factor in this was the attacks by neurologists on asylum-based psychiatry as backward and unscientific, the most famous of which was that by S. Weir Mitchell. For a description of this conflict, see Bonnie Ellen Bluestein, " 'A Hollow Square of Psychological Science': American Neurologists and Psychiatrists in Conflict," in *Madhouses, Mad-Doctors and Madmen: The Social History of Psychiatry in the Victorian Era*, ed. Andrew Scull (Philadelphia: University of Pennsylvania Press, 1981). The rise in prestige and authority of clinical medicine around the turn of the twentieth century is, of course, a complex story involving much more than advances in understanding disease pathogenesis and therapeutics. See Paul Starr, *The Social Transformation of American Medicine* (New York: Basic Books, 1982).

20. Ackerknecht, *Short History of Psychiatry*, p. 66. Austrian psychiatrist Julius Wagner von Jauregg eventually developed an effective therapy for general paresis that involved infecting patients with malaria to induce fever; he was awarded a Nobel Prize for this work in 1927. Ibid., pp. 89–90.

21. For a recent paean to the brilliance of Kraepelin, Alzheimer, and Nissl, see Gayatyri Devi and Wolfgang Quitschke, "Alois Alzheimer, Neuroscientist (1864–1915)," *Alzheimer Disease and Associated Disorders* 13 (1999): 132–137.

22. On Kraepelin's importance to the subsequent history of psychiatry, see Acker-knecht, *Short History of Psychiatry*, pp. 66–71, and Shorter, *History of Psychiatry*, pp. 99–109. Shorter in particular describes the importance of Kraepelin's textbook, which went through eight editions and served as the foundation for modern psychiatric classification (e.g., the American Psychiatric Association's *Diagnostic and Statistical Manual of Mental Disorders*) through the remainder of the twentieth century. Kraepelin's creation of the entity of Alzheimer's disease has spawned a debate that, while lively, is based on very little historical evidence.

23. The clearest description of the ongoing confusion over these issues from Alzhei-mer's time through the 1980s can be found in Thomas Beach, "The History of Alzheimer's Disease: Three Debates," *Journal of the History of Medicine and Allied Sciences* 42 (1987): 327–349.

24. Although this is a fair summary of the conventional account of Alzheimer's report, the case is, of course, a good deal more complicated. For an interesting discussion of the epistemological conundrums entailed in Alzheimer's description of this case and Kraepe-lin's decision to declare it a distinct disease entity, see Rob Dillmann, "Alzheimer Disease: Epistemological Lessons from History?" in *Concepts of Alzheimer Disease: Biological, Clini-cal, and Cultural Perspectives,* ed. Peter J. Whitehouse, Konrad Maurer, and Jesse F. Bal-lenger (Baltimore: Johns Hopkins University Press, 2000). Konrad Maurer and colleagues in Frankfurt recovered Alzheimer's file of this case in the 1990s. Luigi Amaducci and colleagues questioned whether this first case in fact met the criteria for a diagnosis of Alzheimer's disease, arguing that the case as reported more closely resembled another rare form of dementia. Luigi Amaducci et al., "The First Alzheimer Disease Case: A Meta-chromatic Leukodystrophy?" *Developmental Neuroscience* 13 (1991): 186–187. The recovery of Alzheimer's original case file (Konrad Maurer et al., "Auguste D. and Alzheimer's Disease," *Lancet* 349 [1997]: 1546–1549) and this patient's brain sections, which were reexamined (R. B. Graeber et al., "Histopathology and APOE Genotype of the First Alzhei-mer Disease Patient, Auguste D.," *Neurogenetics* 1 [1998]: 223–228), have disproved this hypothesis.

On the development of the silver-staining technique that Alzheimer used, see Heiko Braak and Eva Braak, "Neurofibrillary Changes: The Hallmark of Alzheimer Disease," in Whitehouse, Maurer, and Ballenger, *Concepts of Alzheimer Disease.*

25. Alois Alzheimer, "Über eine eigenartige Erkrankung der Hirnrinde," translated by Katherine Bick as "A Characteristic Disease of the Neocortex," in *The Early Story of Alzhei-mer's Disease: Translation of the Historical Papers by Alois Alzheimer, Oskar Fischer, Emil Kraepelin, Gaetano Perusini,* ed. Katherine Bick, Luigi Amaducci, and Giancarlo Pepeu (Padua, Italy: Liviana Press, 1987).

26. Berrios, "Alzheimer's Disease." Regarding the neurofibrillary tangles, Alzheimer is usually credited with the first observation of them. But Berrios points out that Alzhei-mer's American student Solomon C. Fuller reported their existence some five months before Alzheimer—in an article published in 1907, the same year that Alzheimer published

his case (p. 359). Fuller's article was "A Study of the Neurofibrils in Dementia Paralytica, Dementia Senilis, Chronic Alcoholism, Cerebral Lues and Microencephalic Idiocy," *American Journal of Insanity* 63 (1907): 415–468.

27. Alois Alzheimer, "Über eigenartige Krankheitsfälle des späteren Alters," translated by Hans Förstl and Raymond Levy as "On Certain Peculiar Diseases of Old Age," *History of Psychiatry* 2 (1991): 93; emphasis in original.

28. Berrios, "Alzheimer's Disease," p. 360.

29. Richard M. Torack, "Adult Dementia: History, Biopsy, Pathology," *Neurosurgery* 4 (1979): 432–434.

30. Luigi A. Amaducci, Walter A. Rocca, and Bruce S. Schoenberg, "Origin of the Distinction between Alzheimer's Disease and Senile Dementia: How History Can Clarify Nosology," *Neurology* 36 (1986):1497–1499.

31. Myfanwy Thomas and Michael Isaac, "Alois Alzheimer: A Memoir," *Trends in Neuroscience* 10 (1987): 306–307.

32. Beach, "History of Alzheimer's Disease," pp. 338–341.

33. Berrios, "Alzheimer's Disease," pp. 362–363.

34. Hans Förstl reports a somewhat different dynamic in Germany, where Alzheimer's disease was also accepted as a distinct entity in the second and third decades of the twentieth century. According to him, neuropathological research in this period steadily revealed that the clinical and pathological boundaries of dementia were not clear-cut but shaded from normal to pathological with some overlap in both the pre-senile and senile period. Needing a label to describe the spectrum, researchers in this field used "Alzheimer's disease" because "senile dementia" made no sense in describing dementia that developed in patients as early as their late forties. In other words, Förstl claims that German psychiatrists in that period were using the term as it came to be used in other national contexts only in the 1970s. Hans Förstl, "Contributions of German Neuroscience to the Concept of Alzheimer's Disease," in Whitehouse, Maurer, and Ballenger, *Concepts of Alzheimer Disease.*

35. Holstein also claims that William Malamud studied with Alzheimer, but I have found no source corroborating this.

36. For Holstein's discussion of the attacks from neurology as a factor in accepting Alzheimer's disease as an entity, see pp. 132–137 in her dissertation, "Negotiating Disease"; for her discussion of the researchers training with Alzheimer, see p. 138. For her idea about the importance of the logic of age classification, see Martha Holstein, "Alzheimer's Disease and Senile Dementia, 1885–1920: An Interpretive History of Disease Negotiation." *Journal of Aging Studies* 11 (1997): 1–13. In this article, Holstein argues that another reason American psychiatrists accepted Alzheimer's disease as a legitimate entity was that it allowed them to avoid the therapeutic nihilism associated with the aged. "Researchers could accept brain pathology as the immediate cause of symptoms. Research could seek to modify or even prevent the formation of tangles and plaques" (p. 8). I find this argument untenable. I have not found any evidence that any psychiatrist of the period was considering the possibility—even the distant possibility—of intervening in the formation of plaques and tangles; nor does Holstein provide citations to support the point.

37. Fuller's publications on Alzheimer's disease in the first two decades of the twen-

tieth century were not included, for example, in the collection of historical papers *The Early Story of Alzheimer's Disease* (see n. 25 above), even though his 1907 publication "A Study of the Neurofibrils in Dementia Paralytica, Dementia Senilis, Chronic Alcoholism, Cerebral Lues and Microcephalic Idiocy" was a key precursor to Alzheimer's description of the neurofibrillary tangle and his articles on Alzheimer's disease are crucial to an understanding of the acceptance of it as a disease entity. For the two 1912 articles on Alzheimer's, see notes 39 and 40 in this chapter.

38. M. Kaplan and A. R. Henderson, "Solomon Carter Fuller, M.D. (1872–1953): American Pioneer in Alzheimer's Disease Research," *Journal of the History of the Neurosciences* 9 (2000): 250–261.

39. Solomon C. Fuller, "Alzheimer's Disease (Senium Præcox): The Report of a Case and Review of Published Cases," *Journal of Nervous and Mental Disease* 39 (1912): 440–441.

40. Solomon C. Fuller and Henry I. Klopp, "Further Observations on Alzheimer's Disease," *American Journal of Insanity* 69 (1912): 26, 27.

41. William Malamud and Konstantin Lowenburg, "Alzheimer's Disease: A Contribution to Its Etiology and Classification," *Archives of Neurology and Psychiatry* 21 (1929): 805.

42. David Rothschild and Jacob Kasanin, "Clinicopathologic Study of Alzheimer's Disease: A Contribution to Its Etiology and Classification," *Archives of Neurology and Psychiatry* 36 (1936): 293–94.

43. For other examples of work treating the issue this way, see K. Lowenburg and D. Rothschild, "Alzheimer's Disease: Its Occurrence on the Basis of a Variety of Etiologic Factors," *American Journal of Psychiatry* 11 (1931): 269; K. Lowenburg and R. W. Waggoner, "Familial Organic Psychosis (Alzheimer's Type)," *Archives of Neurology and Psychiatry* 31 (1934): 737; Jacob Kasanin and R. P. Crank, "Alzheimer's Disease," *Annals of Neurology and Psychiatry* 30 (1933): 1180; Alfred Gordon, "Pick's or Alzheimer's Disease," *Archives of Neurology and Psychiatry* 24 (1935): 214; and Alfred Paul Bay and Jack Weinberg, "Review of the Symptomatology of Alzheimer's Disease," *Archives of Neurology and Psychiatry* 47 (1941): 862.

44. Alzheimer, "On Certain Peculiar Diseases of Old Age," p. 93.

45. Fuller, "Alzheimer's Disease (Senium Præcox)," pp. 440, 441.

46. These are briefly described in Pollen's *Hannah's Heirs*, pp. 56, 58. Since Pollen is interested *only* in work that proved important to contemporary research agendas, he ignores the body of work described in the remainder of this chapter.

47. In addition to Beach, "History of Alzheimer's Disease," see Rob Dillmann, *Alzheimer's Disease: The Concept of Disease and the Construction of Medical Knowledge* (Amsterdam: Thesis Publishers, 1990), pp. 161–163, for a discussion of the overuse of the diagnosis of cerebral arteriosclerosis. See Carole Haber, "Geriatrics: A Specialty in Search of Specialists," in Van Tassel and Stearns, *Old Age in a Bureaucratic Society*, for a discussion of the origin and development of the idea that "a man is as old as his arteries."

48. Beach's discussion of this in "History of Alzheimer's Disease" is excellent, as is Haber's in "Geriatrics." For a discussion of recent controversy regarding the aging/disease issue, see Jaber F. Gubrium's *Oldtimers and Alzheimer's: The Descriptive Organization of Senility* (Greenwich, Conn.: JAI Press, 1986), chaps. 1–3. For a recent collection of work on the issue of aging versus disease, see Felicia Huppert, Carol Brayne, and Daniel W.

O'Connor, eds., *Dementia and Normal Aging* (Cambridge: Cambridge University Press, 1994).

49. Lowenburg and Rothschild, "Alzheimer's Disease," 269. Rothschild and Kasanin, "Clinicopathologic Study of Alzheimer's Disease."

50. For a variety of etiologic factors, see Leo Alexander, "Neurofibrils in Systemic Disease and in Supravital Experiments," *Archives of Neurology and Psychiatry* 32 (1934): 293, and Leo Alexander and Joseph M. Looney, "Histologic Changes in Senile Dementia and Related Conditions," *Archives of Neurology and Psychiatry* 40 (1938): 1075. For single cause, see Theodore L. L. Sonati, "Histogenesis of Senile Plaques," *Archives of Neurology and Psychiatry* 46 (1941): 101.

51. Rothschild and Kasanin, "Clinicopathologic Study of Alzheimer's Disease," p. 320. Of course, one might also argue that the difference in clinical pictures between senility and Alzheimer's disease lies more in differential expectations of observers for patients at different ages; that is, because expectations for the emotional functioning of an older person are lower than those for a younger one, an equally devastating clinical picture might be judged more severe in the case of a younger patient.

52. N. Gellerstedt, "Zur Kenntnis der Hirnveränderungen bei der normalen Altersinvolution," *Upsala Läkareförenings Förhandlingar* 38 (1933): 193–408. This article has not been translated into English, but Förstl describes it as a study of the brains of fifty elderly people without dementia, 84% of whom had plaques, the density of which did not appear related to the patients' mental condition. Localization of neurofibrillary tangles appeared to be of greater importance: tangles were localized in the brains of those without dementia but more widely dispersed in the brains of those with senile dementia. Förstl argues that the implication of Gellerstedt's work was that the difference between senile dementia and normal aging was a matter of degree, while the "overstated or simplified" reading of his paper was that brain pathology made no clinical difference. Förstl, "Contributions of German Neuroscience." Although Rothschild never dismissed pathology entirely, some of his followers in the 1940s and 1950s certainly did, as described in the next chapter.

Rothschild's major publications on the psychodynamic approach to senile dementia were David Rothschild, "Pathologic Changes in Senile Psychoses and Their Psychobiologic Significance," *American Journal of Psychiatry* 93 (1937): 757–787; Rothschild and M. L. Sharp, "The Origin of Senile Psychoses: Neuropathologic Factors and Factors of a More Personal Nature," *Diseases of the Nervous System* 2 (1941): 49–54; Rothschild, "Neuropathologic Changes in Arteriosclerotic Psychoses and Their Psychiatric Significance," *Annals of Neurology and Psychiatry* 48 (1942): 417–436; Rothschild, "The Clinical Differentiation of Senile and Arteriosclerotic Psychoses," *American Journal of Psychiatry* 98 (1942): 324; Rothschild, "The Role of the Premorbid Personality in Arteriosclerotic Psychoses," *American Journal of Psychiatry* 100 (1944): 501–505; Rothschild, "Senile Psychoses with Cerebral Arteriosclerosis," in *Mental Disorders in Later Life*, ed. Oscar J. Kaplan (Stanford: Stanford University Press, 1945); Rothschild, "The Practical Value of Research in the Psychoses of Later Life," *Diseases of the Nervous System* 8 (1947): 123; and Sidney L. Sands and Rothschild, "Sociopsychiatric Foundations for a Theory of the Reactions to Aging," *Journal of Nervous and Mental Diseases* 116 (1952): 233–241.

53. Rothschild, "Pathologic Changes in Senile Psychoses," p. 777.

54. Rothschild and Sharp, "Origin of Senile Psychoses," p. 49.

55. Ibid., p. 53.

56. Ibid., p. 54.

57. For a discussion of Rothschild's place in the psychodynamic psychiatry of the period, and particularly the influence on him of Meyer's psychobiology, see Holstein, "Negotiating Disease," pp. 238–245. See Shorter, *History of Psychiatry*, pp. 109–112, for a discussion of Meyer's prominence in American psychiatry, which in Shorter's view was unfortunate. For a thorough discussion on the dominance of psychodynamic psychiatry during this period, see Grob, *Mental Illness and American Society* and *From Asylum to Community*, and Nathan Hale, *The Rise and Crisis of Psychoanalysis in the United States* (New York: Oxford University Press, 1995).

58. Rothschild, "Pathologic Changes in Senile Psychoses," p. 781.

59. Rothschild, "Practical Value of Research in the Psychoses of Later Life," p. 127.

60. Ruth Ehrenberg and Miles O. J. Gullingsrud, "Electroconvulsive Therapy in Elderly Patients," *American Journal of Psychiatry* 111 (1955): 743.

61. G. Wilse Robinson, "The Toxic Delirious Reactions of Old Age," *American Journal of Psychiatry* 99 (1942): 110; Loren Avery et al., "Common Factors Precipitating Mental Symptoms in the Aged," *Archives of Neurology and Psychiatry* 54 (1945): 312; Genevieve Arneson, "Hazards in Tranquilizing the Elderly Patient," *American Journal of Psychiatry* 115 (1958): 163; James Titchener et al., "Psychological Reactions of the Aged in Surgery: The Reaction of Renewal and Depletion," *Archives of Neurology and Psychiatry* 78 (1958): 63.

62. In Avery et al., "Common Factors Precipitating Mental Symptoms in the Aged," pp. 312, 313.

63. Ehrenberg and Gullingsrud, "Electroconvulsive Therapy."

64. Besides Ehrenberg and Gullingsrud, see Vernon L. Evans, "Physical Risks in Convulsive Shock Therapy," *Archives of Neurology and Psychiatry* 48 (1942): 1017, and "Convulsive Shock Therapy in Elderly Patients: Risks and Results," *American Journal of Psychiatry* 99 (1943): 531; Samuel Susselman et al., "Electric Shock Therapy of Elderly Patients," *Archives of Neurology and Psychiatry* 56 (1946): 158; Curtis Prout and Donald Hamilton, "Results of Electroshock Therapy in Patients over Sixty Years of Age," *Archives of Neurology and Psychiatry* 67 (1952): 689; and Gunther E. Wolff, "Results of Four Years Active Therapy for Chronic Mental Patients and the Value of and Individual Maintenance Dose of ECT," *American Journal of Psychiatry*, 114 (1957): 453.

65. Maurice Linden, "Group Psychotherapy with Institutionalized Senile Women: Studies in Gerontologic Human Relations," *Archives of Neurology and Psychiatry* 69 (1953): 400; Raphael Ginzberg, "Geriatric Ward Psychiatry: Techniques in the Psychological Management of Elderly Psychotics," *American Journal of Psychiatry* 110 (1953): 296; Maurice Linden and Douglas Courtney, "Interdisciplinary Research in the Use of Oral Pentylenetetetetrazol (Metrazol) in the Psychoses of Senility and Cerebral Arteriosclerosis," *Archives of Neurology and Psychiatry* 72 (1954): 385; Leopold Judah et al., "Psychiatric Response of Geriatric-Psychiatric Patients to Mellaril," *American Journal of Psychiatry* 115 (1959): 118.

66. Rupert Chittick and Elmer Stotz, "Nicotinic Acid and Ascorbic Acid in Relation to the Care of the Aged," *Diseases of the Nervous System* 2 (1941): 71; W. W. Jetter et al., "Vitamin Studies in Cerebral Arteriosclerosis," *Diseases of the Nervous System* 2 (1941): 66; G. L.

Wadsworth et al., "An Evaluation of Treatment for Senile Psychosis with Vitamin B Complex," *American Journal of Psychiatry* 99 (1943): 807; P. E. Vernon and M. McKinlay, "Effects of Vitamin and Hormone Treatment on Senile Patients," *Journal of Neurology, Neurosurgery, and Psychiatry* 8 (1947): 87; Joseph Haber, "Stellate Ganglion Infiltration in Organic Psychoses of Later Life," *American Journal of Psychiatry* 111 (1955): 751; Ewen Cameron et al., "Interthecal Administration of Hyaluroniadase: Effects upon the Behavior of Patients Suffering from Senile and Arteriosclerotic Behavior Disorders," *American Journal of Psychiatry* 113 (1957): 893; Ewen Cameron, "The Use of Nucleic Acid in Aged Patients with Memory Impairments," *American Journal of Psychiatry* 114 (1958): 943.

67. Jack Pressman, *Last Resort: Psychosurgery and the Limits of Medicine* (Cambridge: Cambridge University Press, 1998); Joel Braslow, *Mental Ills and Bodily Cures: Psychiatric Treatment in the First Half of the Twentieth Century* (Berkeley: University of California Press, 1997).

68. Braslow, *Mental Ills and Bodily Cures*, p. 5. Pressman did not formally elaborate on this distinction but approaches the history of lobotomy from the same contextualist perspective as Braslow does.

69. This generalization is difficult to make because relatively few articles addressed the etiological issue directly. Of the thirty-six articles appearing in the *American Journal of Psychiatry* or the *Archives of Neurology and Psychiatry* between 1940 and 1959 which dealt with etiology and classification, seventeen can easily be classified as definitively taking one position or the other. Of these, twelve adhere to the psychodynamic approach, eleven of which explicitly cite Rothschild as the authority on the subject. Of the five that clearly favor a strict organic explanation, only Neumann and Cohn engage Rothschild's evidence directly. The remaining nineteen articles could not be categorized because their subjects were too narrow. For example, a study ruling out heart disease as a significant factor in the psychoses of the senium could be consistent with either a psychodynamic or an organic model. Of these nineteen uncategorized articles, eleven were concerned with organic issues, five with social issues (three remain sui generis). If these are added in on this basis, the picture is much more balanced—seventeen articles approaching senile dementia from a psychodynamic angle, sixteen from an organic.

70. Meta Neumann and Robert Cohn, "Incidence of Alzheimer's Disease in a Large Mental Hospital: Relation to Senile Psychosis and Psychosis with Cerebral Arteriosclerosis," *Archives of Neurology and Psychiatry* 69 (1953): 615. This single article by Neumann and Cohn perfectly anticipated the position of researchers in the neurosciences who rediscovered senile dementia in the late 1960s. Thus researchers of today interested in tracing their intellectual lineage have repeatedly cited this single article as virtually the only meaningful contribution of American psychiatry to understanding senile dementia during this era.

71. Felix Post, "Some Problems Arising from a Study of Mental Patients over the Age of 60 Years," *Journal of Mental Science* 90 (1944): 564.

72. William H. McMenemey, "Alzheimer's Disease: A Report of Six Cases," *Journal of Neurology and Psychiatry* 3 (1940): 234. For another example of this sort of gloss of Rothschild's work, see James W. Affleck, "Psychiatric Disorders among the Chronic Sick in Hospital," *Journal of Mental Science* 94 (1948): 33.

73. R. D. Newton, "The Identity of Alzheimer's Disease and Senile Dementia and Their Relationship to Senility," *Journal of Mental Science* 94 (1948): 225–248.

74. William Mayer-Gross, "Arteriosclerotic, Senile and Presenile Psychoses," *Journal of Mental Science* 90 (1944): 319.

75. Robert Katzman and Katherine Bick, *Alzheimer Disease: The Changing View* (San Diego: Academic Press, 2000), pp. 46–47.

76. Martin Roth, "The Natural History of Mental Disorders in Old Age," *Journal of Mental Science* 101 (1955): 281–301. A preliminary report on the first 150 patients was published in Roth and J. B. Morrissey, "Problems in the Diagnosis and Classification of Mental Disorder in Old Age, with a Study of Case Material," *Journal of Mental Science* 98 (1952): 66–80.

77. Roth and Morrissey, "Problems in the Diagnosis and Classification of Mental Disorder in Old Age," pp. 77–78.

CHAPTER 3: FROM SENILITY TO SUCCESSFUL AGING

1. Much of my analysis follows from Richard B. Calhoun's *In Search of the New Old: Redefining Old Age in America, 1945–1970* (New York: Elsevier, 1978). Although he ignores the problematic aspects of gerontology's claims to expertise over aging, Calhoun's book easily remains the most thorough description of post–World War II gerontological theory, practice, and advocacy.

More recently, two works have appeared providing important perspectives on the emergence of gerontology: W. Andrew Achenbaum, *Crossing Frontiers: Gerontology Emerges as a Science* (New York: Cambridge University Press, 1995), and Stephen Katz, *Disciplining Old Age: The Formation of Gerontological Knowledge* (Charlottesville: University Press of Virginia, 1996). From a concern with understanding gerontology's status as a profession and a domain of knowledge, Achenbaum argues that, though gerontology has become a well-defined field of scientific inquiry for various disciplines, it has failed to establish the institutional hallmarks of a discipline. From a Foucauldian perspective, Katz argues that gerontology's lack of coherent boundaries makes it a perfect example of a discipline—discursively insinuating itself into every aspect of the lives of the elderly.

2. My use of the word *persuasion* is intended to distinguish the subject matter of this chapter—the core assumptions shared by these diverse groups, their basic way of thinking—from both formal gerontological theory and the concrete policies it advocated. Calhoun's book *In Search of the New Old*, whose subject includes gerontological theory, clinical and social work practice, and public policy, appropriately uses the broader term *gerontological movement*.

3. Achenbaum, *Crossing Frontiers*, pp. 63–68, 75–79, 89–90.

4. Phoebe Liebig, "Professional Organizations," in vol. 3 of the *Encyclopedia of Aging*, ed. David J. Ekerdt (New York: Macmillan Reference USA, 2002), pp. 1133–1139, accessed via *Gale Virtual Reference Library*, Thomson Gale, CIC Penn State University, 18 January 2005, http://find.galegroup.com/gvrl/infomark.do?&type=retrieve&tabID=T001&prod Id=GVRL&docId=CX3402200332&source=gale&userGroupName=psucic& version =1.0.

5. Nathan W. Shock, *Trends in Gerontology,* 2nd ed. (Stanford, Calif.: Stanford University Press, 1957), pp. 139–172.

6. Achenbaum, *Crossing Frontiers,* p. 97.

7. These numbers were cited in Diana Woodruff, ed., *Aging: Scientific Perspectives and Social Issues* (New York: D. Van Nostrand, 1975). The bibliography citing 50,000 publications on aging between 1954 and 1974 was noted in Woodruff's introduction (p. 3). The claim that more gerontological literature had been published between 1950 and 1960 appeared in James E. Birren and Vivian Clayton's chapter in the book, "History of Gerontology," p. 24.

8. This phrase is taken from Chris Gilleard, one of a handful of gerontologists in the 1990s who, as I discuss in chapter 6, call for a return to a psychosocial etiology of senile dementia. Chris Gilleard, "Losing One's Mind and Losing One's Place: A Psychosocial Model of Dementia," in *Gerontology: Responding to an Aging Society,* ed. Kevin Morgan (London: Jessica Kingsley and British Society of Gerontology, 1992).

9. Wilfred M. McClay, *The Masterless: Self and Society in Modern America* (Chapel Hill: University of North Carolina Press, 1994), chap. 7, quotation on p. 234. McClay points out that the popularity of these works can be seen as evidence not only that individual autonomy was threatened by these developments but that Americans continued to prize the ideal of individual autonomy and self-creation. "The social critics were not only prophets of totalist doom; they were also guardians of the self, of its integrity, autonomy, and resiliency" (p. 268).

10. Although Gumpert made no significant contributions to gerontological research, as a popular writer he was one of the most visible proponents of the gerontological persuasion. A measure of Gumpert's prominence in the post–World War II years was that the *New Yorker* ran a two-part profile of him in 1950. D. Lang, "Profiles: Geriatrician, Martin Gumpert," *New Yorker,* 10 June 1950, pp. 30–34; 17 June 1950, pp. 34–36.

11. Martin Gumpert, *You Are Younger Than You Think* (New York: Duell, Sloan and Pierce, 1944), pp. 13–14.

12. Peter Laslett, *A Fresh Map of Life: The Emergence of the Third Age* (Cambridge: Harvard University Press, 1991).

13. These transformations and the ideology that accompanied them are the subject of Calhoun's *In Search of the New Old.* Calhoun's assessment of them as unalloyed progress I find problematic. For more nuanced accounts, see Carole Haber and Brian Gratton's *Old Age and the Search for Security: An American Social History* (Bloomington: Indiana University Press, 1994). Each chapter of the book contains a description of these transformations under the heading "The Social Security Era," and Thomas Cole critiques these transformations in the final chapter of *The Journey of Life: A Cultural History of Aging in America* (Cambridge: Cambridge University Press, 1992).

14. Quotation from Clark Tibbitts and Henry D. Sheldon, "Introduction: A Philosophy of Aging," *Annals of the American Academy of Political and Social Science* 279 (1952): 1–10; this volume is a special issue titled "Social Contributions by the Aging."

15. Several popular medical manuals and scores of popular magazine articles by pioneer geriatricians spread the good news on aging during the 1940s and 1950s. See, for

example, Edward J. Stieglitz's popular book *The Second Forty Years* (1946; facsimile reprint, New York: Arno, 1979) and Gumpert, *You Are Younger Than You Think.*

16. Gumpert, *You Are Younger Than You Think*, p. 13.

17. Leo W. Simmons, *The Role of the Aged in Primitive Society* (New Haven: Yale University Press, 1945). On the book's influence on gerontology, see Calhoun, *In Search of the New Old*, p. 75.

18. Leo W. Simmons, "Social Participation of the Aged in Different Cultures," *Annals of the Academy of Political and Social Science* 279 (1952): 43.

19. Ibid., p. 44.

20. Ibid., pp. 50–51.

21. David Rothschild, "The Practical Value of Research in the Psychoses of Later Life," *Diseases of the Nervous System* 8 (1947): 125.

22. For examples, see Harold W. Williams et al., "Studies in Senile and Arteriosclerotic Psychoses: Relative Significance of Extrinsic Factors in Their Development," *American Journal of Psychiatry* 98 (1942): 712; E. M. Gruenberg, "Community Conditions and Psychoses of the Elderly," *American Journal of Psychiatry* 110 (1954): 888; Ewald W. Busse et al., "Studies of the Process of Aging: Factors That Influence the Psyche of Elderly Persons," *American Journal of Psychiatry* 110 (1954): 897; and Carol Buck, "Environmental Change and Age of Onset of Psychosis in Elderly Patients," *Archives of Neurology and Psychiatry* 75 (1956): 622. The obvious problem with these studies was that they measured only admission rates, not prevalence rates. All these factors could feasibly produce more admissions regardless of whether they actually produced more psychotics.

23. Maurice Linden and Douglas Courtney, "The Human Life Cycle and Its Interruptions: A Psychologic Hypothesis," *American Journal of Psychiatry* 109 (1953): 912.

24. David C. Wilson, "The Pathology of Senility," *American Journal of Psychiatry* 111 (1955): 905.

25. As early as 1934, Rothschild observed in his survey of the literature that the incidence of Alzheimer's disease was 1½ times greater in women than in men, though he did not explicitly discuss this difference. David Rothschild, "Alzheimer's Disease: A Clinico-pathologic Study of Five Cases," *American Journal of Psychiatry* 91 (1934): 485–519. A decade later, William Mayer-Gross noted a higher incidence of reported cases of Alzheimer's disease in his review of the literature and noted that it was "a completely unexplained feature of Alzheimer's disease." William Mayer-Gross, "Arteriosclerotic, Senile and Presenile Psychoses," *Journal of Mental Science* 90 (1944): 316–327. R. D. Newton's 1948 study of 150 autopsied cases found a strikingly greater amount of senile plaques in the brains of women than in those of men—four times greater in the senile (65+) age group and six times greater in the pre-senile (50–64) age group. Newton developed a genetic explanation for this difference. If plaque formation was triggered by a genetic mutation, it would tend to be less prevalent in males because harmful genetic mutations tend to be eliminated by natural selection during the reproductive years, and males have a longer potential reproductive life. R. D. Newton, "The Identity of Alzheimer's Disease and Senile Dementia and Their Relationship to Senility," *Journal of Mental Science* 94 (1948): 225–249. In discussing this work, Daniel Pollen notes that "these observations remain unexplained and

unexploited even to the present day." Daniel Pollen, *Hannah's Heirs: The Quest for the Genetic Origins of Alzheimer's Disease,* expanded ed. (Oxford: Oxford University Press, 1996), p. 58.

26. David D. Sonecypher, "Old Age Need Not Be 'Old,' " *New York Times Magazine,* 18 August 1957, pp. 27, 67.

27. David Rothschild, "The Origin of Senile Psychoses: Neuropathologic Factors and Factors of a More Personal Nature," *Diseases of the Nervous System* 2 (1941): 52.

28. Stonecypher, "Old Age Need Not Be 'Old,' " p. 27. I have found no contemporary evidence in the literature to support Stonecypher's claim here, though it certainly fit the reigning theory well. Recent studies have shown that highly educated people have a lower prevalence of Alzheimer's disease.

29. Martin Gumpert, *Anatomy of Happiness* (New York: McGraw-Hill, 1951), p. 190.

30. Wilma Donahue, "An Experiment in the Restoration and Preservation of Personality in the Aged," in *Planning the Later Years,* ed. Wilma Donahue and Clark Tibbitts (Ann Arbor: University of Michigan Press, 1950), p. 170.

31. Christopher Lasch, *The Culture of Narcissism: American Life in an Age of Diminishing Expectations* (New York: Norton, 1979), p. 207.

32. Stonecypher, "Old Age Need Not Be 'Old,' " p. 27; emphasis in original.

33. Ibid., pp. 27, 67.

34. Ibid., pp. 67–68.

35. Gumpert, *You Are Younger Than You Think,* pp. 62–63.

36. Robert J. Havighurst, "Social and Psychological Needs of the Aging," *Annals of the American Academy of Political and Social Science* 279 (1952): 16.

37. Ibid.

38. Gumpert, *Anatomy of Happiness,* p. 188

39. Gumpert, *You Are Younger Than You Think,* pp. 145, 147–148.

40. Harlow Shapley, "A Design for Fighting," *American Scholar* 14 (1945): 25. Shapley was the director of the Harvard College Observatory.

41. Havighurst, "Social and Psychological Needs of the Aging," p. 17.

42. Eugene A. Friedmann and Robert J. Havighurst, *The Meaning of Work and Retirement* (1954; facsimile reprint, New York: Arno Press, 1977), p. 192; emphasis in original.

43. Ibid.

44. Ibid., p. 194.

45. Jerome Kaplan, *A Social Program for Older People* (Minneapolis: University of Minnesota Press, 1953), pp. 3, 12. Kaplan cites statistics on the overrepresentation of elderly in mental hospitals and provides anecdotal evidence that the existence of community centers and recreation programs drastically lowered the frequency of such hospitalizations among their clients—though he acknowledges that these may be highly selective populations. Surprisingly, I have not found any sociological studies that tried to test this oft made claim.

46. George Gleason, *Horizons for Older People* (New York: Macmillan, 1956), p. 14. Although the comparison to juvenile delinquency was rare, assertions that recreation programs for the elderly lowered the frequency of placement in mental hospitals and other institutions were commonplace. For other examples, see Arline Britton Boucher and John Leo Tehan, "No One under Sixty Need Apply," *Recreation* 42 (1948): 350; Georgene E.

Bowen, "The Time of Their Lives," *Recreation* 44 (1950): 375; and James H. Woods, *Helping Older People Enjoy Life* (New York: Harper and Brothers, 1953), p. 101.

47. Quoted in Gleason, *Horizons for Older people*, p. 7.

48. Edward A. Connell, "How I Want to Grow Old," *America* 99 (1958): 61.

49. Elroy D. Munck, " 'Oldsters' . . . on a Two-Way Street," *Recreation* 52 (1959): 172. See also Carol Lucas, "Antidote for Tarnish," *Recreation* 54 (1961): 514–515.

50. Woods, *Helping Older People Enjoy Life*, pp. 71–72.

51. Carol Lucas, "Not the Same Old Story: A New Frontier for the Aging," *Recreation* 54 (1961): 52.

52. Munck, " 'Oldsters' . . . on a Two-Way Street," pp. 172–173.

53. The concept of the "gerontological web" is developed by Katz in his *Disciplining Old Age*, pp. 111–119.

54. James H. Woods, "How to Tell a Good Golden-Age Club," *Recreation* 46 (1953): 522.

55. Woods, *Helping Older People Enjoy Life*, p. 24; emphasis in original.

56. Ibid., pp. 39–43.

57. Robert N. Butler, *Why Survive? Being Old in America* (New York: Harper and Row, 1975), p. 12. This definition originally appeared in Robert N. Butler and Myrna I. Lewis, *Aging and Mental Health: Positive Psychosocial Approaches* (St. Louis, Mo.: C. V. Mosby, 1973). Butler originally coined the term in a *Washington Post* interview in 1968.

58. Butler, *Why Survive?* pp. 12–13.

59. The scholarly touchstone for gerontologists who saw the aged as a minority group was a 1945 essay by Louis Wirth. Wirth's essay was particularly useful for gerontologists because it defined *minority* not in terms of race, ethnicity, religion, or national identity (though he acknowledged, of course, that these were the categories around which minority group identity was typically constructed). Rather, Wirth defined minority groups in functional terms as "a group of people who, because of their physical or cultural characteristics, are singled out from the others in the society in which they live for differential and unequal treatment, and who therefore regard themselves as objects of collective discrimination." Although Wirth did not specifically say so, his definition made it possible to think of minority group identification as based on a variety of characteristics, such as gender, sexual orientation, or age. Louis Wirth, "The Problem of Minority Groups," in *The Science of Man in World Crisis*, ed. Ralph Linton (New York: Columbia University Press, 1945), p. 347.

For work developing the theoretical framework of the aged as a quasi-minority group, see, for example, Milton Barron, "Minority Group Characteristics of the Aged in American Society," *Journal of Gerontology* 8 (1953): 477–482; Leonard Z. Breen, "The Aging Individual," in *Handbook of Social Gerontology*, ed. Clark Tibbitts (Chicago: University of Chicago Press, 1960); and Irving Rosow, "Old Age: One Moral Dilemma of an Affluent Society," *Gerontologist* 2 (1962): 182–191.

60. Jack Levin and William C. Levin, *Ageism: Prejudice and Discrimination against the Elderly* (Belmont, Calif.: Wadsworth, 1980), pp. 106–107.

61. Arnold M. Rose, "The Subculture of the Aging: A Topic for Sociological Research," *Gerontologist* 2 (1962): 127.

62. Gordon Streib, "Are the Aged a Minority Group?" in *Applied Sociology: Opportunities and Problems*, ed. Alvin W. Gouldner and S. M. Miller (New York: Free Press, 1965).

63. For examples defending the minority group theory, see Erdman Palmore and Frank Whittington, "Trends in the Relative Status of the Aged," *Social Forces* 50 (1971): 84-91, and Levin and Levin, *Ageism.*

64. Butler, *Why Survive?* p. 322.

65. Ibid., 355.

66. Margaret E. Kuhn, "What Older People Want for Themselves and for Others in Society," in *Advocacy and Age: Issues, Experiences, Strategies,* ed. Paul A. Kerschner (Los Angeles: University of Southern California Press, 1976).

67. Alex Comfort, *A Good Age* (New York: Simon and Schuster, 1976), pp. 63-64.

68. Ibid., p. 23.

69. Kenneth Stampp, *The Peculiar Institution: Slavery in the Ante-Bellum South* (New York: Vintage, 1956), pp. vii-ix.

70. Daniel P. Moynihan, *The Negro Family: The Case for National Action* (Washington, D.C.: GPO, 1969). As described by historian William Tratner, the Moynihan report and similar work "posited the existence of a subculture with patterns of behavior that distinguished it from the larger social structure and prevented its members—who exhibited such aberrant psychological and moral traits and values as feelings of fatalism, helplessness, inferiority, dependence, and present-mindedness—from taking advantage of the opportunities available to better their lives." Although Moynihan intended the opposite, conservatives used this idea to claim that those perpetuating this "culture of poverty" were undeserving of public assistance. William Tratner, *From Poor Law to Welfare State: A History of Social Welfare in America,* 5th ed. (New York: Free Press, 1994), p. 330.

71. Cole, *Journey of Life,* p. 229.

72. Comfort, *Good Age,* p. 47.

CHAPTER 4: THE RENAISSANCE OF PATHOLOGY

1. J. A. N. Corsellis, *Mental Illness and the Aging Brain: The Distribution of Pathological Change in a Mental Hospital Population* (London: Oxford University Press, 1962), p. 2.

2. Organic mental disorders are characterized by the presence of physiological or anatomical changes to the nervous system, functional disorders by their absence. Since the functional disorders, by definition, were not characterized by any identifiable brain pathology, the brains of patients who had developed such disorders at any age could serve as controls to measure the relative importance of pathological change in the organic disorders.

3. Corsellis, pp. 57-59.

4. This is implicitly evident in all the Newcastle group's publications but made explicit in the introduction to Bernard E. Tomlinson, Gary Blessed, and Martin Roth, "Observations on the Brains of Non-Demented Old People," *Journal of Neurological Science* 7 (1968): 331: "Whether these degenerative changes are solely responsible for the clinical syndrome of dementia has been questioned, mainly on the grounds that similar changes may be found in mentally 'normal' old people. . . . Rothschild (1937, 1942, 1956) found a lack of correlation between the degree of intellectual impairment and the histological changes of senile psychosis and between the mental changes and the extent of the brain destruction in arteriosclerotic psychosis; he expressed the view that degenerative changes within the brain

were less important than the individual's response to the changes and that a degree of cerebral degeneration which produces dementia in one person may be well tolerated by another. This lack of correlation between the degenerative changes and mental function has been restated by Wolf (1959), and Noyes and Kolb (1963) emphasize the interaction of organic and psychological factors in the development of senile psychosis."

5. The clinical and pathological procedures employed by the group to derive quantitative data are thoroughly described in Gary Blessed, Bernard E. Tomlinson, and Martin Roth, "The Association between Quantitative Measures of Dementia and of Senile Change in the Cerebral Gray Matter of Elderly Subjects," *British Journal of Psychiatry* 114 (1968): 797–811. The dementia scale and tests are included in an appendix to the article (pp. 808–809).

To my knowledge, no researcher has attempted to generate a data set that is *directly* comparable to the Newcastle data, an indication of the complexities involved in this study despite its apparent simplicity. Comparative studies have been made, but they have involved refinements in both the methods of clinical and pathological evaluation and indication and so have generated data that are not directly comparable.

6. Martin Roth, Bernard E. Tomlinson, and Gary Blessed, "Correlations between Scores for Dementia and Counts of 'Senile Plaques' in Cerebral Grey Matter of Elderly Patients," *Nature* 209 (1966): 110. A fuller version of this brief report appeared in Roth, Tomlinson, and Blessed, "The Relationship between Quantitative Measures of Dementia and of Degenerative Changes in the Cerebral Grey Matter of Elderly Subjects," *Proceedings of the Royal Society of Medicine* 60 (1967): 254–260.

7. Blessed, Tomlinson, and Roth, "Association between Quantitative Measures of Dementia and of Senile Change," pp. 801–808.

8. Bernard E. Tomlinson, Gary Blessed, and Martin Roth, "Observations on the Brains of Demented Old People," *Journal of Neurological Science* 11 (1970): 234–235.

9. Blessed, Tomlinson, and Roth, "Association between Quantitative Measures of Dementia and of Senile Change," p. 807.

10. Ibid., p. 805.

11. Tom Kitwood, "Explaining Senile Dementia: The Limits of Neuropathological Research," *Free Associations* 10 (1987): 124–125.

12. For my discussion of the consistency of this finding since Rothschild's time, see note 25 in chapter 3.

13. Eighteen percent of the cases involved both vascular and Alzheimer-type lesions, and the remainder had other diagnoses or could not be determined.

14. For assessments of the importance of the Newcastle study to the subsequent development of Alzheimer's disease research, see W. A. Lishman, "The History of Research into Dementia and Its Relationship to Current Concepts," in *Dementia and Normal Aging*, ed. Felicia Huppert, Carol Brayne, and Daniel W. O'Connor (Cambridge: Cambridge University Press, 1994), and Robert Katzman and Katherine L. Bick, "The Rediscovery of Alzheimer's Disease in the Decades of the 1960s and 1970s," in *Concepts of Alzheimer Disease: Biological, Clinical and Cultural Perspectives*, ed. Peter J. Whitehouse, Konrad Maurer, and Jesse F. Ballenger (Baltimore: Johns Hopkins University Press, 2000).

15. For example, in his influential article "From Senility to Alzheimer's Disease: The Rise of the Alzheimer's Disease Movement" (*Milbank Memorial Fund Quarterly* 67, no. 1

(1989): 64), Patrick Fox suggested that the elimination of the distinction between Alzheimer's disease and senile dementia "was primarily due to the application of newly developing technologies, principally the electron microscope, to the study of neurological diseases."

16. Michael Kidd, "Paired Helical Filaments in Electron Microscopy of Alzheimer's Disease," *Nature* 197 (1963): 192–193, and "Alzheimer's Disease: An Electron Microscopic Study," *Brain* 86 (1964): 309–310, for his description of the paired helical filament.

17. Robert Terry, "The Fine Structure of Neurofibrillary Tangles in Alzheimer's Disease," *Journal of Neuropathology and Experimental Pathology* 22 (1963): 629–642; Robert Terry, Nicholas K. Gonatas, and Martin Weiss, "Ultrastructural Studies in Alzheimer's Presenile Dementia," *American Journal of Pathology* 44 (1964): 269–281.

18. In G. E. W. Wolstenholme and Maeve O'Connor, *Alzheimer's Disease and Related Conditions: A Ciba Foundation Symposium* (London: J&A Churchill, 1970).

19. Henryk M. Wisniewski, H. K. Narang, and Robert D. Terry, "Neurofibrillary Tangles of Paired Helical Filaments," *Journal of Neurological Science* 27 (1976): 173–181.

20. In describing Divry's work, I follow the discussion of Robert D. Terry and Henrik M. Wisniewski in "The Ultrastructure of the Neurofibrillary Tangle and the Senile Plaque," in Wolstenholme and O'Connor, *Alzheimer's Disease and Related Conditions*. For another assessment of the importance of Divry's work, see Pollen, *Hannah's Heirs*, p. 58.

21. Terry and Wisniewski, "Ultrastructure of the Neurofibrillary Tangle and the Senile Plaque," p. 159.

22. Quotation found in the discussion of the paper by Dorys Hollander and Sabina J. Strich, "Atypical Alzheimer's Disease with Congophilic Angiopathy, Presenting with Dementia of Acute Onset," in Wolstenholme and O'Connor, *Alzheimer's Disease and Related Conditions*, p. 125.

23. Terry and Wisniewski, "Ultrastructure of the Neurofibrillary Tangle and the Senile Plaque," p. 162. Terry elaborates on this hypothesis in the discussion of the article: "I think the single primary structural abnormality in this disease is the twisted tubule, which possibly gives rise to all the other changes, perhaps even changing axoplasmic flow" (p. 165).

24. In Wolstenholme and O'Connor, *Alzheimer's Disease and Related Conditions*, p. 166.

25. Katzman and Bick ("Rediscovery of Alzheimer's Disease") characterize it as the first modern symposium on Alzheimer's disease. It brought together the key players in the 1960s who were determined to develop a biological approach to Alzheimer's disease, and it was the first high-level conference organized around Alzheimer's disease and related disorders as discrete entities distinct from aging.

26. Martin Roth, "Chairman's Closing Remarks," in Wolstenholme and O'Connor, *Alzheimer's Disease and Related Conditions*, p. 303.

27. Ibid., p. 305.

28. Robert Terry, "Dementia: A Brief and Selective Review," *Archives of Neurology* 33 (1976): 3.

29. On rare occasions, references to the possibility of a psychosocial etiology for Alzheimer-type dementia appeared in the literature. For example, Raymond T. Bartus and colleagues noted that "although sociocultural, economic and psychological factors probably

contribute to the cognitive deterioration, the medical community commonly believes that age-related dysfunctions in the central nervous system are intimately involved." The authors made no further comment on this and include no references to the early literature. Raymond T. Bartus et al., "The Cholinergic Hypothesis of Geriatric Memory Dysfunction," *Science* 217 (1982): 408. Even rarer was research seriously treating psychosocial factors as a possible etiology of dementia. One example is an article by Barry J. Gurland in a book of conference proceedings. Citing Rothschild, Gurland argued that "it is still an open matter whether there is an important sociocultural contribution to the prevalence of Alzheimer's and related forms of dementia occurring in the senium, but . . . the evidence now available is sufficiently intriguing to warrant further study of the issue." In the discussion following the article, Martin Roth, not surprisingly, treated the argument derisively, suggesting that such claims needed to be based on research as complete as the Newcastle study before they deserved serious consideration. Barry J. Gurland, "The Borderlands of Dementia: The Influence of Sociocultural Characteristics on the Rates of Dementia Occurring in the Senium," in *Clinical Aspects of Alzheimer's Disease and Senile Dementia*, ed. Nancy E. Miller and Gene D. Cohen (New York: Raven Press, 1981).

30. Debate about the relative importance of the two in the pathogenesis of Alzheimer's disease has persisted to the present. After the precise proteins for plaques (beta amyloid) and tangles (tau) were identified, this disagreement has been described as the conflict between the "baptists" (beta amyloid) and the "taoists" (tau).

31. Rob Dillmann makes this case in chapter 6 of *Alzheimer's Disease: The Concept of Disease and the Construction of Medical Knowledge* (Amsterdam: Thesis Publishers, 1990).

32. A. H. Pope et al., "Microchemical Pathology of the Cerebral Cortex in Presenile Dementias," *Transactions of the American Neurological Association* 89 (1965): 15–16.

33. A brief summary of these precursors of the cholinergic hypothesis may be found in Raymond T. Bartus et al., "The Cholinergic Hypothesis: A Historical Overview, Current Perspectives and Future Directions," *Annals of the New York Academy of Science* 444 (1986): 333. A more detailed discussion can be found in Dillmann, *Alzheimer's Disease*, pp. 210–225.

34. D. A. Drachman and J. Leavitt, "Human Memory and the Cholinergic System: A Relationship to Aging?" *Archives of Neurology* 30 (1974): 113–121.

35. Dillmann, *Alzheimer's Disease*, p. 220.

36. Elaine Perry's group also suggested in passing that obtaining biopsy material from patients with dementia was unethical, which raises interesting questions about the common use of biopsy tissue by the electron microscopists. (The representativeness of the tissue was not an issue for the electron microscopists, since they were examining the ultrastructure of various pathological features.) Perry and colleagues raise the question with the clause "even if biopsy were ethical in dementia in old age." Perry et al., "Neurotransmitter Enzyme Abnormalities in Senile Dementia," *Journal of the Neurological Sciences* 34 (1977): 248. The basis for this objection is not explicated in the article—that is, whether biopsy is unethical because of the problem of obtaining informed consent from patients with dementia or because the procedure is unjustifiably dangerous or destructive to the patient; nor are there any citations to discussion of cerebral biopsy as an ethical issue.

37. Perry et al., "Neurotransmitter Enzyme Abnormalities in Senile Dementia," pro-

vides both a concise summary and fairly detailed discussion of these problems. See also Elaine K. Perry et al., "Circadian Variations in Cholinergic Enzymes and Muscarinic Receptor Binding Activities in Human Cerebral Cortex," *Neuroscience Letters* 4 (1977): 185–189; Robert H. Perry et al., "Human Brain Temperature at Necropsy—A Guide in Post-mortem Biochemistry," *Lancet* 1 (1977): 38; David M. Bowen et al., "Chemical Pathology of the Organic Dementias: Validity of Biochemical Measurements on Human Post-Mortem Brain Specimens," *Brain* 100 (1977): 397–426; and Peter Davies and A. J. F. Maloney, "Selective Loss of Cholinergic Neurons in Alzheimer's Disease," *Lancet* 2 (1976): 1403. Each of these researchers makes a short list of potential sources of artifactual contamination and the steps taken to minimize them.

38. Dillmann, *Alzheimer's Disease*.

39. Katzman and Bick, "Rediscovery of Alzheimer's Disease," provides a few biographical details on Davies, Bowen, and Perry.

40. Katzman and Bick, "Rediscovery of Alzheimer's Disease."

41. Elaine K. Perry et al., "Necropsy Evidence of Central Cholinergic Deficits in Senile Dementia," *Lancet* 1 (1977): 189. The group concludes its report in *Neuroscience Letters* 6 (1977): 85–89, with a statement of the clinical relevance of the cholinergic hypothesis as well.

42. Elaine K. Perry et al., "Correlation of Cholinergic Abnormalities with Senile Plaques and Mental Test Scores in Senile Dementia," *British Medical Journal* 2 (1978): 1457–1459.

43. Whitehouse and colleagues describe these studies in their initial report linking the nucleus basalis of Meynert with the cholinergic deficit. Peter Whitehouse et al., "Alzheimer's Disease: Evidence for Selective Loss of Cholinergic Neurons in the Nucleus Basalis," *Annals of Neurology* 10 (1981): 122.

44. Ibid., p. 123.

45. Peter Whitehouse et al., "Alzheimer's Disease and Senile Dementia: Loss of Neurons in the Basal Forebrain," *Science* 215 (1982): 1237–1239.

46. Ibid., p. 1239.

47. Quoted in Gina Kolata, "Clues to Alzheimer's Disease Emerge," *Science* 219 (1983): 941–942.

48. For a summary of the neurotransmitter alterations found in Alzheimer's disease, see Joseph Coyle et al., "Alzheimer's Disease: A Disorder of Cortical Cholinergic Innervation," *Science* 219 (1983): 1184–1190.

49. Quoted in Kolata, "Clues to Alzheimer's Disease Emerge," p. 942.

50. Coyle et al., "Alzheimer's Disease," p. 1188; emphasis added.

51. Perry et al., "Correlation of Cholinergic Abnormalities with Senile Plaques and Mental Test Scores in Senile Dementia."

52. The studies by Drachman showed that performance on memory tests was most clearly affected, while performance on tests of other cognitive abilities, for example, verbal IQ, was relatively unimpaired. Animal studies seemed to be constructed to measure the effect of lesions on memory. Drachman and Leavitt, "Human Memory and the Cholinergic System."

53. Peter Whitehouse et al., "Dementia: Bridging the Brain-Behavior Barrier," in *Senile*

Dementia of the Alzheimer Type, ed. J. T. Hutton and A. D. Kenny (New York: Alan Liss, 1985).

54. Bartus et al., "Cholinergic Hypothesis: A Historical Overview," p. 332.

55. David A. Drachman, "The Cholinergic System, Memory and Aging," in *Brain Neurotransmitters and Receptors in Aging and Age-Related Disorders,* ed. S. J. Enna et al. (New York: Raven Press, 1981).

56. Whitehouse et al., "Alzheimer Disease: Evidence for Selective Loss of Cholinergic Neurons in the Nucleus Basalis."

57. Richard J. Wurtman, "Alzheimer's Disease," *Scientific American,* January 1985, p. 62.

58. For a review essay outlining understanding in the 1970s of mild age-associated memory impairments, see V. A. Kral, "Benign Senile Forgetfulness," in *Alzheimer's Disease: Senile Dementia and Related Disorders,* ed. Robert Katzman, Robert D. Terry, and Katherine L. Bick (New York: Raven Press, 1978).

59. Whitehouse et al., "Dementia," pp. 225–226.

60. Peter Whitehouse, "Clinical Trials in Alzheimer's Disease: State of the Art and Future Directions," in *Alzheimer's Disease: New Treatment Strategies,* ed. Zhaven S. Khachaturian and John P. Blass (New York: M. Dekker, 1992).

61. Bartus et al., "Cholinergic Hypothesis: A Historical Overview," pp. 341–343, for a discussion of early therapeutic efforts.

62. More than a dozen drugs were in various stages of development as of 1994. Ezio Giacobini and Robert Becker, "Development of Drugs for Alzheimer Therapy: A Decade of6Progress," in *Alzheimer Disease: Therapeutic Strategies,* ed. Ezio Giacobini and Robert Becker (Boston: Birkhauser, 1994).

63. Gina Bari Kolata, "Clues to the Cause of Senile Dementia: Patients with Alzheimer's Disease Seem to be Deficient in a Brain Neurotransmitter," *Science* 211 (1981): 1032–1033.

64. Felicia Huppert and Carol Brayne, "What Is the Relationship between Dementia and Normal Aging?" in Huppert, Brayne, and O'Connor, *Dementia and Normal Aging.* The implication of a gene on chromosome 21 for familial/early-onset cases suggests that this relatively small number of cases at least ought to be regarded as a distinct entity. But even here, the variability of time in which individuals carrying the genetic mutation develop dementia suggests that other factors are at play, including the effects of aging. See S. J. Richards and C. Van Broeckhoven, "Genetic Linkage in Alzheimer's Disease," in the same volume.

65. In Wolstenholme and O'Connor, *Alzheimer's Disease and Related Conditions,* p. 284.

66. Katzman, Terry, and Bick, *Alzheimer's Disease,* pp. 265–266.

67. Robert Katzman, "The Prevalence and Malignancy of Alzheimer Disease: A Major Killer," *Archives of Neurology* 33 (1976): 217–218.

68. Katzman and Bick, "Rediscovery of Alzheimer's Disease."

69. Robert Katzman, Robert D. Terry, and Katherine Bick, "Recommendations of the Nosology, Epidemiology, and Etiology and Pathophysiology Commissions of the Workshop-Conference on Alzheimer's Disease–Senile Dementia and Related Disorders," in Katzman, Terry, and Bick, *Alzheimer's Disease,* quotation from general discussion on p. 268.

70. V. C. Hachinski and N. A. Lassen, "Multi-Infarct Dementia: A Cause of Mental Deterioration in the Elderly," *Lancet* 2 (1974): 207–209.

71. National Institute on Aging Consensus Task Force, "Senility Reconsidered: Treatment Possibilities for Mental Impairment in the Elderly," *Journal of the American Medical Association* 244 (1980): 259–263.

72. Guy McKhann et al., "Clinical Diagnosis of Alzheimer's Disease: Report on the NINCDS-ADRDA Work Group under the Auspice of the Department of Health and Human Services Task Force on Alzheimer's Disease," *Neurology* 34 (1984): 939–944.

73. Alzheimer's Disease Education and Referral Center, National Institute on Aging, "Alzheimer's Disease Fact Sheet," July 2004.

74. Marshall F. Folstein and Paul R. McHugh, "Dementia Syndrome of Depression," in Katzman, Terry, and Bick, *Alzheimer's Disease*, p. 92.

75. Katzman, Terry, and Bick, *Alzheimer's Disease*, p. 94.

76. Charles E. Wells, "Pseudodementia," *American Journal of Psychiatry* 136 (1979): 897.

77. McKhann et al., "Clinical Diagnosis of Alzheimer's Disease," p. 940.

78. J. Grimley Evans, "Ageing and Disease," in *Research and the Ageing Population: CIBA Foundation Symposium 134*, ed. David Evered and Julie Whelan (New York: John Wiley and Sons, 1988). The exchange between Katzman and Evans appears in the discussion following Evans' paper, p. 47.

79. James Goodwin, "Geriatric Ideology: The Myth of the Myth of Senility," *Journal of the American Geriatrics Society* 39 (1991): 627–628.

80. As I discuss in the introduction and in chapter 6, British psychologist Tom Kitwood, who developed a comprehensive psychosocial model of the etiology of Alzheimer's disease and geared a treatment program around that model, has been harshly critical of the ideological component of biomedical research, arguing that it has resulted in the widespread acceptance of a flimsy disease construct. Kitwood's critique of biomedical research on dementia is especially prominent in "Explaining Senile Dementia"; "Towards the Reconstruction of an Organic Mental Disorder," in *Worlds of Illness*, ed. Alan Radley (London: Routledge, 1993); and "Dementia: Social Section, Part II," in *History of Clinical Psychiatry: The Origin and History of Psychiatric Disorders*, ed. German E. Berrios and Roy Porter (New York: New York University Press, 1995). Other critics, such as Patrick Fox, Karen Lyman, and Ann Robertson, tend to allow that scientific explanations are legitimate within the constructed sphere of scientific discourse (thus glossing over the complexities and contingencies I have described in this chapter), while criticizing science for its reductionist tendency to exclude the social. Although all these critiques have their merits, they fail to consider the ways in which similar ideological commitments structure psychosocial explanations. Moreover, they fail to recognize the ways in which both biomedical science and its critics share a common ideological orientation toward old age. See Fox, "From Senility to Alzheimer's Disease"; Karen Lyman, "Bringing the Social Back In: A Critique of the Biomedicalization of Dementia," *Gerontologist* 29 (1989): 597–605; and Ann Robertson, "The Politics of Alzheimer's Disease: A Case Study in Apocalyptic Demography," in *Critical Perspectives on Aging*, ed. Meredith Minkler and Carroll L. Estes (Amityville, N.Y.: Baywood, 1991).

CHAPTER 5: THE HEALTH POLITICS OF ANGUISH

1. U.S. Congress, *Congressional Record* (Washington, D.C.: GPO), pp. 24327–24329.

2. From the complete text of the letter, found at www.americanpresidents.org/letters/ 39.asp, accessed 2 February 2005.

3. See the Alzheimer's Association Web site for a news release outlining the Reagan family's commitment to the fight against Alzheimer's disease: www.alz.org/Media/news releases/ronaldreagan/reagan family.asp, accessed 1 February 2005.

4. Data for 1976 and 1983 taken from Patrick Fox, "From Senility to Alzheimer's Disease: The Rise of the Alzheimer's Disease Movement," *Milbank Memorial Fund Quarterly* 67, no. 1 (1989): 96; data for 1994 from the Alzheimer's Association's 1994 Annual Report, p. 14; data for 2005 from a news release on the Alzheimer's Association Web site, www.alz.org/Media/newsreleases/2004/112304_congress.asp, accessed 25 May 2005.

5. So far as I can tell, the term *Alzheimer's disease movement* originated with Fox's 1989 article "From Senility to Alzheimer's Disease," and not from Alzheimer's disease advocates themselves. The title of Fox's article notwithstanding, it is essentially a study of the Alzheimer's Association. He makes little effort to discern the degree to which there was organized social action related to Alzheimer's disease *not* connected with the association. The definition of a social movement he employs ("a set of opinions and beliefs in a population representing preferences for changing some elements of the social structure or reward distribution, or both, of a society" [p. 65]) is so broad that virtually any organization could be included. As a result, for him the history of the association *is* the history of the movement.

6. This argument is only implied in the 1989 article. He states it explicitly in Patrick Fox, "The Role of the Concept of Alzheimer's Disease," in *Concepts of Alzheimer Disease: Biological, Clinical, and Cultural Perspectives,* ed. Peter J. Whitehouse, Konrad Maurer, and Jesse F. Ballenger (Baltimore: Johns Hopkins University Press, 2000).

7. Fox, "From Senility to Alzheimer's Disease," pp. 79–80.

8. Ibid., pp. 75–82.

9. The scientific content of this milestone conference is described in detail in chapter 4.

10. J. Mitchell to R. Katzman, 15 July 1980, quoted in Fox, "From Senility to Alzheimer's Disease," pp. 85–86.

11. The groups from Seattle and San Francisco were the ones that withdrew; the Pittsburgh group was the third dissenting group. Fox, "From Senility to Alzheimer's Disease," p. 85.

12. Ibid., p. 82.

13. Fox, "Role of the Concept of Alzheimer Disease."

14. In discussing the failed struggle for long-term care insurance, Robert Binstock calls the association's efforts "particularly noteworthy." Robert Binstock, "The Politics of Enacting Long-Term Care Insurance," in *The Future of Long-Term Care: Social and Policy Issues,* ed. Robert Binstock, Leighton E. Cluff, and Otto von Mering (Baltimore: Johns Hopkins University Press, 1996), p. 221.

15. Although Alzheimer's disease advocates headed off this challenge through the 1980s, it was made in the 1990s. Richard C. Adelman argued that the NIA spent a dispro-

portionately large share of its resources on Alzheimer's research at the expense of other important research in gerontology and basic science. Richard C. Adelman, "The Alzheimerization of Aging," *Gerontologist* 35 (1995): 526–532.

16. Lewis Thomas, "The Problem of Dementia," 1981, reprinted in *Late Night Thoughts on Listening to Mahler's Ninth Symphony* (New York: Viking, 1983), p. 121.

17. Stone arrived at this figure by taking a frequently cited estimate of the total cost of nursing home care related to Alzheimer's disease—$13 billion—and adding to it his estimate of the costs of providing care for people with Alzheimer's disease living at home; this second estimate was based on his own experience and that of other caregivers in the ADRDA times the estimated number of individuals with Alzheimer's disease living at home—which totaled another $13 billion.

18. U.S. House of Representatives, Select Committee on Aging, 98th Cong., 1st sess., *Senility: The Last Stereotype* (Washington, D.C.: GPO, 1983), p. 41 (hereafter cited as "U.S. House, *Senility*"). Stone made this case in his columns in the ADRDA's first two annual reports as well.

19. *ADRDA 1984 Annual Report*, p. 2. A comparison of the amount spent on care versus the amount spent on research appeared in his column in the 1985 *Annual Report* as well.

20. David A. Drachman, "Drachman Summarizes Mini–White House Proceedings," *ADRDA Newsletter* 1, no. 1 (1981): 6; emphasis in original.

21. Lewis Thomas, "The Technology of Medicine," in *Lives of a Cell* (New York: Viking, 1974), pp. 31–35.

22. Ibid., p. 36.

23. Louise Detenbeck, letter to the editor, *ADRDA Newsletter* 7, no. 2 (1987): 7. The article in question was "ADRDA, NIA Mobilize to Study THA: Putting 'Breakthroughs' in Perspective," *ADRDA Newsletter* 7, no. 1 (1987): 1.

24. U.S. House, *Senility*, pp. 59–60.

25. U.S. Senate, Subcommittee on Aging, Committee on Labor and Human Resources, 96th Cong., 2nd sess., *Impact of Alzheimer's Disease on the Nation's Elderly* (Washington, D.C.: GPO, 1980), p. 116 (hereafter cited as "U.S. Senate, *Impact of Alzheimer's Disease on the Nation's Elderly*").

26. U.S. Senate, Special Committee on Aging, 98th Cong., 1st sess., *Endless Night, Endless Mourning: Living with Alzheimer's* (Washington, D.C.: GPO, 1983), p. 48 (hereafter cited as "U.S. Senate, *Endless Night*").

27. Ibid., p. 54.

28. U.S. House of Representatives, Select Committee on Aging, 99th Cong., 1st sess., *Alzheimer's Disease: Burdens and Problems for Victims and their Families* (Washington, D.C.: GPO, 1986), p. 24 (hereafter cited as "U.S. House, *Alzheimer's Disease*").

29. U.S. Senate, *Endless Night*, pp. 1–2.

30. Ibid., pp. 3–4.

31. Ibid., p. 10.

32. U.S. House, *Alzheimer's Disease*, p. 42.

33. Katherine L. Bick, "The History of the Alzheimer's Association: Future Public Policy Implications," in Whitehouse, Maurer, and Ballenger, *Concepts of Alzheimer Disease*.

34. U.S. House, *Senility*, p. 90.

35. Ibid., p. 91

36. Ibid., p. 21.

37. As I have argued throughout this book, stereotypes are more than simple misrepresentations based on ignorance or hostility (or both). They are one of the fundamental mechanisms people employ to organize experience. As such, stereotypes are inevitable—though the particular shape they take is contingent and variable. Exploring stereotypes provides a way to understand the meaning of a disease by showing us what anxieties are connected with it. Put another way, to understand the meaning of a disease we must understand what people take to be at stake in the distinction between the normal and the pathological which the notion of disease presumes. In this way, stereotypes of disease tell us what a society is most anxious about losing.

38. Jaber F. Gubrium, *Oldtimers and Alzheimer's: The Descriptive Organization of Senility* (Greenwich, Conn.: JAI Press, 1986); Gubrium, "Structuring and Destructuring the Course of Illness: The Alzheimer's Disease Experience," *Sociology of Health and Illness* 9 (1987): 1–24; Gubrium, "Narrative Practice and the Inner Worlds of the Alzheimer Disease Experience," in Whitehouse, Maurer, and Ballenger, *Concepts of Alzheimer Disease;* Gubrium and Robert J. Lynott, "Measurement and the Interpretation of Burden in the Alzheimer's Disease Experience," *Journal of Aging Studies* 1 (1987): 265–85; Gubrium and Lynott, "Alzheimer's Disease as Biographical Work," in *Social Bonds in Later Life,* ed. Jill Quadagno (Beverly Hills, Calif.: Sage, 1985).

39. The Seattle group, the Alzheimer's Support, Information and Service Team, was one of the founding groups of the ADRDA and one of the two that left the association over the issue of the disease-specific focus of the group and the primary emphasis on support for biomedical research which flowed from that focus.

40. U.S. Senate, *Impact of Alzheimer's Disease on the Nation's Elderly,* p. 75.

41. Nancy L. Mace and Peter V. Rabins, *The 36-Hour-Day: A Family Guide to Caring for Persons with Alzheimer's Disease, Related Dementing Ilnesses and Memory Loss in Later Life* (Baltimore: Johns Hopkins University Press, 1981). In 1998, the book entered its third edition.

42. U.S. Senate, *Impact of Alzheimer's Disease on the Nation's Elderly,* p. 98.

43. U.S. Senate, *Endless Night,* p. 12.

44. I base this point on James G. Patterson's *The Dread Disease: Cancer and Modern American Culture* (Cambridge: Harvard University Press, 1987). Although Patterson does not develop the point, he shows (pp. 32–33 and 56–57) that from the late nineteenth century on it was clearly recognized that the incidence of cancer increased with age. But it seems to me equally clear that in the discourse around cancer that he examines throughout the book, younger people—typically parents of young children—were the examples who generated sympathy for the War on Cancer.

45. Even after multiple genetic factors were discovered in the 1990s, increased age remained easily the biggest risk factor for Alzheimer's disease.

46. U.S. House, *Senility,* p. 21.

47. The first discussion I have found in Alzheimer's disease movement literature is an

article by Linda K. George in the ADRDA newsletter: "Adult-Child Caregivers: Caught in the 'Sandwich' of Competing Demands," *ADRDA Newsletter* 7, no. 1 (1987): 8.

48. Stuart Roth, "Reagan's Courage Should Help Others Overcome Fear of Alzheimer's," 10 November 1994, text of op-ed piece distributed to all Alzheimer's Association chapters for placement in local media outlets under Roth's name or the name of the local's president or executive director. Benjamin Green-Field Library, offices of the National Alzheimer's Association, Chicago. Emphasis in original.

49. Matt Clark et al., "A Slow Death of the Mind," *Newsweek*, 3 December 1984, pp. 56–62.

50. Mace and Rabins, *36-Hour Day*, p. xiv.

51. Donna Cohen and Carl Eisdorfer, *The Loss of Self: A Family Resource for the Care of Alzheimer's Disease and Related Disorders* (New York: Norton, 1986).

52. I am not aware of any study that analyzes the ethnic composition of the association's membership, though this is certainly an interesting and relevant question.

53. Mace and Rabins, *36-Hour Day*, p. 7.

54. U.S. Senate, *Endless Night*, pp. 23, 25. The lawyer was Peter Strauss.

55. Ibid., p. 58.

56. Ibid., p. 60.

57. Mace and Rabins, *36-Hour Day*, p. 7.

58. Ibid., p. 9.

59. Jay Mark Ellis, foreword to Jean Tyler (told by Harry Anifantakis), *The Diminished Mind: One Family's Battle with Alzheimer's Disease* (Blue Ridge Summit, Pa.: TAB Books, 1991), p. vii.

60. Of course, this is not to suggest that people who, one way or another, are outside the broad notion of "middle class" do not experience stigma from Alzheimer's disease. Indeed, one would expect that—political representations notwithstanding—factors such as poverty would compound the suffering and stigma of Alzheimer's disease. But this broad notion of "middle class" dominated the Alzheimer's disease movement (as it dominated most other areas of American culture). The experiences of other people have not been represented in the Alzheimer's disease policy discourse, and it would require another sort of study to bring that experience to light.

61. Guy Lushin, *The Living Death: Alzheimer's in America* (n.p.: National Foundation for Medical Research, 1990); emphasis in original.

62. Alzheimer's Disease and Related Disorders Association, graphic with caption "Alzheimer's Disease. The Long Goodbye," *ADRDA Newsletter* 5, no. 4 (1985): back cover.

63. U.S. Senate, *Impact of Alzheimer's Disease on the Nation's Elderly*, p. 2.

64. Suzanne Johnson, "Alzheimer's: Disease of the Unburied Dead," *ADRDA Newsletter* 4, no. 2 (1984): 2.

65. Stacie Leimas, letter to the editor, *ADRDA Newsletter* 7, no. 1 (1987): 7.

66. Sara Townsend, "Grandma," *ADRDA Newsletter* 5, no. 2 (1985): 10.

67. Linda Hubbard, "The Alzheimer Puzzle: Putting the Pieces Together," *Modern Maturity*, August–September 1984, p. 44.

68. Thomas, "Problem of Dementia," p. 121.

69. Clark et al., "Slow Death of the Mind," p. 56.

70. Cohen and Eisdorfer, *Loss of Self*, pp. 22–24.

71. Ibid., p. 143.

72. Mace and Rabins, *36-Hour Day*, p. 164.

73. Cohen and Eisdorfer, *Loss of Self*, pp. 180–181.

74. Ann Wiesner, "My Grandfather Is an Abandoned House," *ADRDA Newsletter* 6, no. 2 (1986): 8.

75. Lushin, *Living Death*, pp. 52, 54.

76. Thomas, "Problem of Dementia," p. 121.

77. NIA, "Senility: Myth or Madness," *NIA Age Page* (Washington, D.C.: GPO, 1980).

78. Robin Marantz Henig, *The Myth of Senility: The Truth about the Brain and Aging*, updated and revised ed. (Glenview, Ill.: Scott, Foresman and Co., 1988), p. xiii. The phrase "the real senile dementia" is used in the title of one of the book's chapters.

79. Gero Vita Laboratories, *Stop the Clock*, advertising brochure (Markham, Ontario, Canada: Gero Vita Laboratories, n.d.).

80. Charles Leroux, "A Silent Epidemic," *Chicago Tribune*, 1981; reprinted by the ADRDA. Benjamin Green-Field Library, Offices of the National Alzheimer's Association, Chicago.

81. Johnson, "Alzheimer's."

82. Frances Guisinger, "Shattered Dreams," *ADRDA Newsletter* 5, no. 1 (1985): 8.

83. U.S. Senate, *Impact of Alzheimer's Disease on the Nation's Elderly*, p. 30.

84. Cohen and Eisdorfer, *Loss of Self*, pp. 58–59.

85. From the reproduction of the note printed in Jack Kevorkian's *Prescription: Medicide. The Goodness of Planned Death* (Buffalo, N.Y.: Prometheus Books, 1991), p. 228. According to Kevorkian's account, by the time she made the trip from Portland to Detroit to obtain his assistance with her suicide, Adkins' dementia had progressed to the point where she had forgotten how to form a cursive capital *A* in her signature on the consent forms—evident in the word "alzheimers" in the suicide note. Her husband, Ronald Adkins, showed her how to make a cursive *A* on another sheet of paper, and she was able to complete her signature. The first news story on the Adkins case in the *New York Times* quoted Ronald Adkins as reading from the note, but the quotation differs in some small ways that do not seem significant from the note reproduced in Kevorkian's book. "Doctor Tells of First Death Using His Suicide Device," *New York Times*, 6 June 1990, p. A1.

86. Peter G. Filene, *In the Arms of Others: A Cultural History of the Right-to-Die Movement* (Chicago: Ivan R. Dee, 1998), chap. 2.

87. Ibid., p. 67. The books are Stanley Keleman's *Living Your Dying* (New York: Random House/Bookworks, 1974) and Marjorie McClay, *To Die with Style* (Nashville: Abingdon Press, 1974).

88. Derek Humphrey, "Rational Suicide among the Elderly," in *Suicide and the Older Adult*, ed. A. A. Leenaars et al. (New York: Guilford, 1992), pp. 125–129. My assessment of Humphrey's article is taken from Stephen Post, "Alzheimer Disease and Physician Assisted Suicide," *Alzheimer Disease and Associated Disorders* 7 (1993): 65–68.

89. This was the chronology reported by the Seattle medical team in its published

account of her case. Kirsten Rohde, Elaine R. Peskind, and Murray R. Raskind, "Suicide in Two Patients with Alzheimer's Disease," *Journal of the American Geriatrics Society* 43 (1995): 187–189.

90. Timothy Egan, " 'Her Mind Was Everything,' Dead Woman's Husband Says," *New York Times*, 6 June 1990, p. B6.

91. Timothy Egan, "As Memory and Music Faded, Oregon Woman Chose Death," *New York Times*, 7 June 1990, p. A1.

92. Michael Betzold, *Appointment with Doctor Death* (Troy, Mich.: Momentum Books, 1993), p. 43.

93. Lawrence K. Altman, "Use of Suicide Device Sets in Motion Debate on a Disturbing Issue," *New York Times*, 12 June 1990, p. C3.

94. Nancy Gibbs, "Dr. Death's Suicide Machine," *Time*, 18 June 1990, pp. 69–70. The expert was Dr. Joanne Lynn, professor at George Washington University, who was a prominent advocate of hospice care. *Newsweek* quoted Stephen Miles, medical ethicist at the University of Minnesota: "Among the things that are lost first in Alzheimer's are insight and memory, both of which are critical to making this decision." "The Doctor's Suicide Van," *Newsweek*, 18 June 1990, p. 47.

95. Quoted in "Doctor's Suicide Van," p. 46.

96. "Dying, Dr. Kevorkian's Way," editorial, *New York Times*, 7 June 1990, p. A22.

97. Marcia Angell, "Don't Criticize Dr. Death . . . ," *New York Times*, 14 June 1990, p. A27.

98. For summaries of the early reaction to the Adkins case by bioethicists and other health policy experts, see Andrew Malcolm, "Giving Death a Hand: A Rending Issue," *New York Times*, 9 June 1990, p. 16, and Altman's "Doctor's World" column "Use of Suicide Device Sets in Motion Debate on a Disturbing Issue."

99. "Doctor's Suicide Van," p. 49.

100. Stephen Post, *The Moral Challenge of Alzheimer's Disease* (Baltimore: Johns Hopkins University Press, 1995).

101. Kevorkian, *Prescription*, p. 222.

102. Stephen G. Post, "The Concept of Alzheimer Disease in a Hypercognitive Culture," in Whitehouse, Maurer, and Ballenger, *Concepts of Alzheimer Disease*.

103. The phrase comes from Laurence B. McCullough's "Ethical Challenges Posed by Injuries and Diseases of the Nervous System," *Medical Humanities Review* 10, no. 2 (1996): 108–112, a harsh review of Post's *Moral Challenge of Alzheimer Disease*.

104. "Oregon Woman Who Chose Death Is Remembered at Upbeat Service," *New York Times*, 11 June 1990, p. D13.

105. Quoted in Betzold, *Appointment with Doctor Death*, p. 48.

CHAPTER 6: THE PRESERVATION OF SELFHOOD IN THE CULTURE OF DEMENTIA

1. Stephen G. Post, "The Concept of Alzheimer Disease in a Hypercognitive Society," in *Concepts of Alzheimer Disease: Biological, Clinical, and Cultural Perspectives*, ed. Peter J. Whitehouse, Konrad Maurer, and Jesse F. Ballenger (Baltimore: Johns Hopkins University Press, 2000).

2. Evidence for this may be found in Jaber F. Gubrium's ethnographic work on Alzheimer's disease caregivers support groups: "Narrative Practice and the Inner Worlds of the Alzheimer Disease Experience," in Whitehouse, Maurer, and Ballenger, *Concepts of Alzheimer Disease.*

3. Arthur W. Frank, *The Wounded Storyteller: Body, Illness, and Ethics* (Chicago: University of Chicago Press, 1995), p. 60.

4. Ibid., chap. 1.

5. For a fuller discussion of this idea, see Jesse F. Ballenger and Peter J. Whitehouse, "The Body as Cultural Text," *Medical Humanities Review* 12, no. 1 (1998): 96–99.

6. Joseph M. Foley, "The Experience of Being Demented," in *Dementia and Aging: Ethics, Values, and Policy Choices,* ed. Robert H. Binstock, Stephen G. Post, and Peter J. Whitehouse (Baltimore: Johns Hopkins University Press, 1992), p. 42.

7. For example, see Victoria Cotrell and Richard Schulz, "The Perspective of the Patient with Alzheimer's Disease: A Neglected Dimension of Dementia Research," *Gerontologist* 33 (1993): 205–211, and Lisa Snyder, *Speaking Our Minds: Personal Reflections from Individuals with Alzheimer's* (New York: W. H. Freeman 1999).

8. Gubrium, "Narrative Practice."

9. Gubrium, in fact, gives no explanation for why individual caregivers and caregiver support groups differ on the issue of whether a person with Alzheimer's disease retains his or her selfhood. Gubrium's project is rooted in a Geertzian anthropology that seeks to demonstrate the ways in which culture is produced locally, in this case, in support groups. He analyzes the way in which the interactions of the members of these groups produce a particular understanding of the subjective world of the person with Alzheimer's disease; while his analysis is insightful and provides a crucial means of resisting overly broad generalizations about the impact of mass culture, this approach has its limits. A satisfying explanation of *why* caregivers have the particular beliefs they do requires an analysis of larger structures of meaning as reflected in admittedly problematic terms I have employed throughout this book—terms such as *society, American culture,* or, in the paragraphs that follow, *spirituality* and *Christianity.*

10. Theresa Strecker, *Alzheimer's: Making Sense of Suffering* (Lafayette, La.: Vital Issues Press, 1997), p. 142.

11. Ibid., pp. 103–104.

12. Ann Davidson, *Alzheimer's: A Love Story: One Year in My Husband's Journey* (Secaucus, N.J.: Birch Lane Press, 1997), pp. 31–32.

13. Ibid.

14. Ibid., p. 223.

15. Lela Knox Shanks, *Your Name Is Hughes Hannibal Shanks: A Caregiver's Guide to Alzheimer's,* 2nd ed. (New York: Penguin, 1999), p. 31.

16. Ibid., pp. 34, 36.

17. Ibid., pp. 34–37.

18. Ibid., pp. 2–4.

19. Ibid., p. 31.

20. Ibid., p. 38.

21. Although Shanks had graduated magna cum laude with a journalism degree, she

did not work outside the home for most of their marriage because Hughes strongly believed that "a family needs someone at home for backup and support at all times." She took a job outside the home only after their children were fully grown and Hughes was nearly set to retire. She retired herself after seven years. Remarkably, she began her career lecturing on "African-American history, how women can achieve wholeness, the development of an inner life, and caregiving" only in 1989, *after* the onset of her husbands dementia. "I had to go to work to help pay for Hughes's medical expenses. I had done public speaking nearly all my life but had never before been paid for it." Ibid., pp. 3–4. For another caregiver account that deftly uses sources from literature, the humanities, and social sciences to critically explore the interrelationships between the person with dementia and his or her family, see Judith Levine's *Do You Remember Me?: A Father, a Daughter, and a Search for Self* (New York: Free Press, 2004).

22. These quotations appear on p. 33 of Shanks' book and come from Robert Bogdan and Steven Taylor, "Relationships with Severely Disabled People: The Social Construction of Humanness," *Social Problems* 36 (1989): 136, 145.

23. Shanks, *Your Name Is Hughes Hannibal Shanks*, p. 95, for a description of Shanks' relationship with the two older women who served as spiritual guides.

24. David Keck, *Forgetting Whose We Are: Alzheimer's Disease and the Love of God* (Nashville, Tenn.: Abingdon Press, 1996), p. 33.

25. Tristram Engelhardt Jr., *The Foundations of Bioethics* (Oxford: Oxford University Press, 1986), p. 109, quoted in David Smith, "Seeing and Knowing Dementia," in Binstock, Post, and Whitehouse, *Dementia and Aging*, p. 47.

26. Smith, "Seeing and Knowing Dementia," p. 52.

27. Stephen Post, *The Moral Challenge of Alzheimer Disease* (Baltimore: Johns Hopkins University Press, 1995), pp. 32–34.

28. Bogdan and Taylor, "Relationships with Severely Disabled People," p. 146. In her sketch of the historical development of theories of self, sociologist Diane Bjorklund cites Cooley's *Human Nature and the Social Order* (1902) and Mead's *Mind, Self and Society* (a volume of posthumously published material produced during Mead's career at the University of Chicago, where he taught from 1900 until his death in 1931) as the beginning point of the theorization of self as a social creation. Diane Bjorklund, *Interpreting the Self: Two Hundred Years of American Autobiography* (Chicago: University of Chicago Press, 1998). Bogdan and Taylor and most of the other authors cited in this section cite Mead, but Erving Goffman's *The Presentation of Self in Everyday Life* (1959) and *Stigma: Notes on the Management of Spoiled Identity* (1963) figure more prominently as the wellspring of contemporary theories of the social construction of the self.

29. Bogdan and Taylor, "Relationships with Severely Disabled People," p. 146.

30. This is not to suggest that there is a necessary opposition between religious belief and various social constructionist approaches. To be sure, reconciling these two ways of knowing is a challenge, but it is far from impossible. While acceptance of a social constructionist position would make it impossible to accept the claim of traditional religious orthodoxies to serve as the transcendent, ahistorical foundation of all human knowledge, there are other approaches to religious belief that allow for historical contingency and development, emphasizing, for example, that God's relationship to humanity is a gradual

process of dialectical unfolding in accord with the particular situations that humans have created for themselves.

31. Bogdan and Taylor, "Relationships with Severely Disabled People," pp. 135–136. The authors are reacting to more than two decades of scholarship that was geared to revealing the ways in which deviancy was managed by society and by those labeled deviant. The inevitability of labeling and stigmatization has become an unexamined axiom in this literature, the authors claim. "The rejection and exclusion of deviant groups are so taken for granted that when labeled deviants are not stigmatized and rejected, such reaction is often described as 'denial' and the 'cult of the stigmatized' " (p. 135). This formulation suggests an interesting parallel with the ways in which "denial" is described in caregiver manuals and sometimes in support groups. Typically, manuals employ some sort of staged model reminiscent of that of Elisabeth Kübler-Ross, whereby caregivers move from anger and denial to understanding and acceptance of the disintegration of their loved ones. Gubrium describes caregiver support groups that, in contrast to the caregiver support groups that encourage their members to see people with dementia as fully human beings, chide caregivers for denial when they refuse to accept that they will eventually have to put their loved one in a nursing home. Gubrium, "Narrative Practice."

32. In addition to the article by Fontana and Smith, the only other constructionist study I am aware of which supports the idea that the self is in fact lost in Alzheimer's disease is that of the Israeli researchers Hava Golander and Aviad E. Raz, "The Mask of Dementia: Images of 'Demented Residents' in a Nursing Ward," *Ageing and Society* 16 (1996): 269–285. Golander and Raz describe the positive and negative social roles created and maintained for patients with dementia by patients without dementia in an Israeli nursing home. The study by Golander and Raz is in general more nuanced than that of Fontana and Smith, avoiding the reiteration of the stock language of the Alzheimer's movement about the "loss of self" in the disease (which could possibly reflect some differences in the Israeli context). They respond directly to the account by Sabat and Harré, arguing that the existence of personal selfhood requires some mastery of social content as well as the employment of first-person indexicals: "First-person indexicals (indeed discourse in general) could be imitated in the formal, syntactically-preserved but content-lacking speech of demented residents. In order to argue that indexicals are 'meaningful,' one should be able to expose their 'intention.' This difficult task becomes highly problematic in the case of dementia" (p. 283).

33. In addition to the account of Sabat and Harré, other works that provide a similar account of the possibilities of reconstructing selfhood in Alzheimer's disease are Chris Gilleard, "Losing One's Mind and Losing One's Place: A Psychosocial Model of Dementia," in *Gerontology: Responding to an Aging Society,* ed. Kevin Morgan (London: Jessica Kingsley and British Society of Gerontology, 1992); Jaber F. Gubrium, "The Social Preservation of Mind: The Alzheimer's Disease Experience," *Symbolic Interaction* 9 (1986): 37–51; Karen Lyman, "Bringing the Social Back In: A Critique of the Biomedicalization of Dementia," *Gerontologist* 29 (1989): 597–605; Tom Kitwood, "Brain, Mind and Dementia: With Particular Reference to Alzheimer's Disease," *Ageing and Society* 9 (1989):1–15; Kitwood, "The Dialectics of Dementia: With Particular Reference to Alzheimer's Disease," *Ageing and Society* 10 (1990): 177–196; Kitwood, *Dementia Reconsidered: The Person Comes First* (Buckingham, England: Open University Press, 1997); and Kitwood and Kathleen Bredin, "To-

wards a Theory of Dementia Care: Personhood and Well-being," *Ageing and Society* 12 (1992): 269–287.

34. George Herbert Mead, *Mind, Self and Society: From the Standpoint of a Social Behaviorist* (1934; Chicago: University of Chicago Press, 1967), pp. 173, 175.

35. Goffman's account of selfhood contains many other elements, but these are not relevant to my discussion here. Erving Goffman, *The Presentation of Self in Everyday Life* (Garden City, N.Y.: Doubleday Anchor, 1959).

36. Ibid., pp. 252–253.

37. Ibid., pp. 253–254.

38. Andrea Fontana and Ronald W. Smith, "Alzheimer's Disease Victims: The 'Unbecoming' of Self and the Normalization of Competence," *Sociological Perspectives* 32 (1989): 45, 39.

39. Ibid., pp. 36, 41.

40. Ibid.; example found on p. 38.

41. Ibid., p. 40.

42. Ibid., pp. 41–43.

43. Their theory of the personal self was based on developments in psycholinguistics, the writings of Wittgenstein, and the developmental psychology of Vygotsky. See Rom Harré, *Personal Being* (Oxford: Blackwell, 1983) and "The Discursive Production of Selves," *Theory and Psychology* 1 (1991): 51–63.

44. Steven Sabat and Rom Harré, "The Construction and Deconstruction of Self in Alzheimer's Disease," *Ageing and Society* 12 (1992): 445–447.

45. Ibid., p. 452.

46. Ibid., p. 450.

47. Ibid., pp. 443–444.

48. Gilleard, "Losing One's Mind and Losing One's Place."

49. Sabat and Harré, "Construction and Deconstruction of Self in Alzheimer's Disease," p. 455.

50. Of course, these accounts do differ in terms of their general characterization of the social interactions between people with dementia and their caregivers, even though their observations are gathered from the same type of institution—an adult day care center. Whereas Fontana and Smith characterize these interactions as generally positive and affirming of the selfhood of people with dementia, Sabat and Harré are more inclined to see them as negative and delegitimating. It is not clear the degree to which this is a difference in evidence, that is, whether the institutions studied are characterized by radically different sorts of social interactions, but it certainly reflects the different interpretive imperatives of the authors, which lead them to make different choices as to what sort of social interactions to emphasize. Moreover, the positive or negative tone ascribed to any interaction is also dependent to a degree on the interpretive position of the authors.

51. Elizabeth Herskovits, "Struggling over Subjectivity: Debates about the 'Self' and Alzheimer's Disease," *Medical Anthropology Quarterly* 9 (1995): 148.

52. Keck, *Forgetting Whose We Are*, p. 32.

53. First-person dementia memoirs are a growing genre. See Thaddeus M. Raushi, *A View from Within: Living with Early Onset Alzheimer's* (Albany, N.Y.: Alzheimer's Associa-

tion, Northeastern New York Chapter, 2001); a second volume of memoirs by Larry Rose, *Larry's Way: Another Look at Alzheimer's from the Inside* (New York: iUniverse, 2003); and two books by Christine Bryden, who was a high-level civil servant in Australia when she was diagnosed with Alzheimer's at the age of 46 (rediagnosed three years later as fronto-temporal dementia) and in 2003 became the first person with dementia to be appointed to the board of directors of a major Alzheimer's disease voluntary association—the London-based Alzheimer's Disease International. Her books are *Dancing with Dementia: My Story of Living Positively with Dementia* (London: Jessica Kingsley, 2005) and *Who Will I Be When I Die?* (East Melbourne: HarperCollins, 1998), the latter published under the name Christine Boden, her name before marriage.

Much of the writing from people in the early stage of dementia has been connected to the organization Dementia Advocacy and Support Network International (DASNI), which developed out of a Yahoo-based discussion group in 2000. The organization advocates for greater understanding and improved services for people with dementia and provides an important forum for sharing experiences. DASNI's first board of directors included Christine Bryden and Jeanne Lee, whose book is described in this chapter. Larry Rose was also a member of the organization, and some of its other members have given presentations at high-level conferences for researchers and professionals working in the Alzheimer's field and have appeared in the mass media. DASNI's Web site is www.dasninternational.org.

54. Peter Whitehouse gently suggests this possibility in his review of DeBaggio's first book, *Losing My Mind*, and criticizes what he sees as its overemphasis on the frightening aspects of the disease experience. *New England Journal of Medicine* 347 (2002): 861.

55. Cary Smith Henderson, *Partial View: An Alzheimer's Journal* (Dallas: Southern Methodist University Press, 1998), p. 4.

56. Frank discusses the quest narrative in chapter 5 of *The Wounded Storyteller*.

57. Robert Davis, *My Journey into Alzheimer's Disease* (Wheaton, Ill.: Tyndale House, 1983), p. 56.

58. Ibid., p. 18.

59. Larry Rose, *Show Me the Way to Go Home* (Forest Knolls, Calif.: Elder Books, 1996), p. 4.

60. Ibid., p. 12.

61. Ibid., pp. 127–128.

62. Diana Friel McGowin, *Living in the Labyrinth: A Personal Journey through the Maze of Alzheimer's* (New York: Delacorte, 1993), p. 97.

63. Ibid., pp. 125–126.

64. *Just Love Me: My Life Turned Upside Down by Alzheimer's* (West Lafayette, Iowa: Purdue University Press, 2003), pp. 23–24.

65. Ibid., pp. 34–35.

66. The books do not make clear exactly what role his wife and son played in the preparation of the books for publication, and there is no discernible change in the quality of the prose within the book or from one volume to the next. In the acknowledgments to the second volume, DeBaggio credits them with having "actively helped in writing this book." Thomas DeBaggio, *When It Gets Dark: An Enlightened Reflection on Life with Alzheimer's* (New York: Free Press, 2003).

67. Thomas DeBaggio, *Losing My Mind: An Intimate Look at Life with Alzheimer's* (New York: Free Press, 2002), p. 160.

68. Ibid., pp. 43, 157.

69. Ibid., p. 29.

70. Ibid., pp. 199, 207.

71. Ibid., pp. 225–226.

72. Henderson, *Partial View,* p. 35.

73. Ibid., p. 4.

74. Ibid., p. 93.

75. Golander and Raz, "Mask of Dementia."

76. Sabat and Harré, "Construction and Deconstruction of Self in Alzheimer's Disease," p. 461.

77. Steven R. Sabat. "Voices of Alzheimer's Disease Sufferers: A Call for Treatment Based on Personhood," *Journal of Clinical Ethics* 9 (1998): 47.

78. Lyman. "Bringing the Social Back In," pp. 597–600.

79. Edmund G. Howe, "Caring for Patients with Dementia: An Indication for 'Emotional Communism,'" *Journal of Clinical Ethics* 9 (1998): 3–11. The special issue had articles by, among others, Kitwood, Lyman, Sabat, and Steven Post.

80. Kitwood, *Dementia Reconsidered,* p. 144.

81. Kitwood discusses Rothschild in "Dementia: Social Section, Part II," in *A History of Clinical Psychiatry: The Origin and History of Psychiatric Disorders,* ed. German E. Berrios and Roy Porter (New York: New York University Press, 1995).

82. Michel Foucault, *Madness and Civilization: A History of Insanity in the Age of Reason* (New York: Random House, 1965).

83. Kitwood and Bredin, "Towards a Theory of Dementia Care," p. 273; emphasis in original.

84. Kitwood, *Dementia Reconsidered,* pp. 5, 144.

85. This point has been articulated by Peter J. Whitehouse and William E. Deal, "Situated beyond Modernity: Lessons for Alzheimer's Research," *Journal of the American Geriatrics Society* 43 (1995): 1314–1315.

86. Kenneth Gergen, *The Saturated Self: Dilemmas of Identity in Contemporary Life* (New York: Basic Books, 1991).

87. Christopher Lasch, *The Minimal Self: Psychic Survival in Troubled Times* (New York: Norton, 1984), pp. 57–58.

88. McGowin, *Living in the Labyrinth,* p. 112.

89. Henderson, *Partial View,* p. 47.

90. Susan Sontag, *Illness as Metaphor* (New York: Vintage, 1979).

91. Kitwood and Bredin perhaps acknowledged the potential for this sort of dynamic in their suggestion that dementia caregivers needed to borrow the principles of self-analysis and awareness articulated in the field of psychotherapy and counseling. In these fields, it is "axiomatic that any person who would help others to deal with their distress and self-defeating patterns must first get acquainted with his or her own personal difficulties." I say "perhaps acknowledged" because, where this line of thought would seem logically to lead them to consider whether their own anxieties encourage them to see the potential for the

person with dementia in perhaps unrealistically positive terms, withal they are inclined to suppose that such processes account only for the negative attitudes that most biomedical researchers and many caregivers exhibit toward people with dementia. Kitwood and Bredin, "Towards a Theory of Dementia Care," p. 274.

92. Rose, *Show Me the Way to Go Home*, pp. 71–72; emphasis in original.

93. Barbara Clow, in "Whose Afraid of Susan Sontag? or, the Myths and Metaphors of Cancer Reconsidered," *Social History of Medicine* 14, no. 2 (2001): 293–312, directly refutes Sontag's claim that until recently the stigmatization of cancer resulted in a shroud of secrecy. Moreover, she showed ways in which patients employed metaphor in a way that was empowering to them. Other historians who have investigated the questions raised by Sontag have tended to reject her central claim that scientific progress in understanding the etiology and the development of effective treatments inevitably frees patients from stigmatizing metaphors and stereotypes. Among the more prominent works are Alan Brandt, *No Magic Bullet: A Social History of Venereal Disease in the United States since 1880*, expanded ed. (New York: Oxford University Press, 1987); Sander L. Gilman, *Difference and Pathology: Stereotypes of Sexuality, Race and Madness* (Ithaca, N.Y.: Cornell University Press, 1985) and *Disease and Representation: Images of Illness from Madness to AIDS* (Ithaca, N.Y.: Cornell University Press, 1988); and Sheila Rothman, *Living in the Shadow of Death: Tuberculosis and the Social Experience of Disease in American History* (New York: Basic Books, 1994). Brandt shows how, despite a thorough understanding of the etiology and the existence of efficacious treatment, venereal disease continues to carry the cultural burden of middle-class sexual anxieties. Gilman argues that all representations of disease, including those of scientific medicine, are ways of managing anxiety about the coherence of the self by creating a stereotypical "Other" beset by disease and dissolution. Most directly, Rothman argues that in the case of tuberculosis, the disease on which Sontag based her claim that stigma is decreased when etiology is understood and effective treatment developed, progress in medical science led to greater stigmatization and marginalization of those who had the disease.

94. On medical diagnosis as value-laden, see Julia Epstein's study of the production of medical case histories, which suggests that there are not essential differences between this work and other literary genres, such as fiction and ethnography, which are supposed to be less objective. Julia Epstein, *Altered Conditions: Disease, Medicine and Storytelling* (New York: Routledge, 1995). From a more conventional position that does not fundamentally challenge the objectivity of the diagnostic process, Howard Brody argues that even in conventional terms to diagnose a disease is to interpret a particular set of signs, to give it a meaning that will have important ramifications for the patient. Howard Brody, *Stories of Sickness* (New Haven, Conn.: Yale University Press, 1987).

95. Frank, *Wounded Storyteller*, p. 66.

Index